SpringerBriefs in Electrical and Computer Engineering

W0079783

More information about this series at http://www.springer.com/series/10059

Lei Yang • Miao He • Junshan Zhang • Vijay Vittal

Spatio-Temporal Data Analytics for Wind Energy Integration

 Springer

Lei Yang
Electrical Computer & Energy Engineering
 Ira A. Fulton School of Engineering
Arizona State University
Tempe, Arizona
USA

Miao He
Department of Electrical and Computer
 Engineering
Texas Tech University
Lubbock, Texas
USA

Junshan Zhang
Electrical Computer & Energy Engineering
 Ira A. Fulton School of Engineering
Arizona State University
Tempe, Arizona
USA

Vijay Vittal
Electrical Computer & Energy Engineering
 Ira A. Fulton School of Engineering
Arizona State University
Tempe, Arizona
USA

ISSN 2191-8112   ISSN 2191-8120 (electronic)
SpringerBriefs in Electrical and Computer Engineering
ISBN 978-3-319-12318-9   ISBN 978-3-319-12319-6 (eBook)
DOI 10.1007/978-3-319-12319-6

Library of Congress Control Number: 2014954748

Springer Cham Heidelberg New York Dordrecht London

Printed on acid-free paper

Springer is part of Springer Science+Business Media (www.springer.com)

Preface

To address the grand challenge of a sustainable energy future, there has been a surge of interest in renewable energy resources, including wind, solar, and bio-fuels. A very recent study by National Renewable Energy Lab (NREL), namely Western Wind and Solar Integration Study Phase 2, determined that up to 33 % wind and solar energy production in the Western grid of the United States would avoid dramatic reduction of carbon dioxide and nitrogen oxides emission. This level of greenhouse gas (GHG) emission reduction points to a bright future for renewable energy resources, provided that advancements are made to adequately integrate these resources in electric grid operations by developing tools and techniques to account for their inherent variability. The high penetration of wind energy generation, with their non-stationary and variable generation characteristics, however, introduces difficult-to-control dynamics and challenges for power system operation.

Building on the insight that massive data collected at wind farms contain rich statistical information, we devise data analytics based stochastic models to significantly improve the forecast accuracy of wind generation and aim to address the controllability and reliability of renewable generation integration. Specifically, in Chap. 1, we give an overview of the state-of-the-art wind power forecast. Chaps. 2 and 3 focus on the development of short-term wind power forecast using a spatio-temporal analysis approach. In Chap. 2, we investigate short-term forecast of wind farm generation by applying spatio-temporal analysis to extensive measurement data collected from a large wind farm and develop finite-state Markov-chain-based forecast models. In Chap. 3, we enhance the finite-state Markov models developed in Chap. 2 by incorporating the wind ramp forecast using support vector machines. In Chap. 4, we investigate stochastic optimization of economic dispatch using the proposed short-term wind power forecast. Finally, we draw conclusions and outline future research directions in Chap. 5. We hope this brief could be a useful reference for graduate students and professionals who are interested in data analytics for wind energy integration in smart grid.

We would like to thank National Renewable Energy Laboratory (NREL) and Xcel Energy for providing the data used in this study. We also would like to thank the Springer editors and staff for their great help in getting this brief published. This research work was supported in part by the US National Science Foundation under

Grant CPS-1035906, in part by the DTRA grant HDTRA1-13-1-0029, and in part by the Power System Engineering Research Center.

Tempe, AZ, USA   Lei Yang
Lubbock, TX, USA   Miao He
Tempe, AZ, USA   Junshan Zhang
Tempe, AZ, USA   Vijay Vittal

Contents

Chapter 1
Introduction

The age of a "smart" grid is upcoming. In the past decades, tremendous national efforts have been dedicated to building the elements of a modern electric power grid, which will ultimately become a smart grid. Futuristic smart grids are envisaged to dramatically increase the efficiency of electricity production and distribution, reduce greenhouse gas emissions and support a sustainable energy infrastructure.

1.1 Data Analytics in Smart Grid

The future smart grid is expected to leverage advanced information and communication technologies (ICTs) to improve grid operations and planning. For example, widely dispersed phasor measurement units (PMUs) and PMU-enabled intelligent electronic devices (IEDs) make wide-area monitoring, protection and control (WAMPAC) possible; smart meters and advanced metering infrastructure (AMI) enable two-way communications between end-users and utilities and sophisticated demand side management; at wind farms, measurement data of individual turbines power output at high temporal resolutions facilitates wind power generation analysis and forecast.

Therefore, future smart grids have the potential to generate massive amounts of data from widely deployed measurement devices. Massive amounts of more detailed data collected from networked measurement devices provides great opportunities for enhanced situational awareness. On the other hand, it also raises new challenges for the effective extraction of relevant information from massive data so as to support decision making. Under increasingly dynamic and uncertain conditions of the power grid, new computational methods are necessary for efficient management of massive data. New algorithms for data fusion, data mining and data analytics have to be developed based on deeper understanding of the spatial and temporal dynamics of power systems.

© The Author(s) 2014
L. Yang et al., *Spatio-Temporal Data Analytics for Wind Energy Integration*,
SpringerBriefs in Electrical and Computer Engineering, DOI 10.1007/978-3-319-12319-6_1

1.2   Impact of Wind Power Integration on Power System

For bulk power systems, reliability is defined by the North American Electric Reliability Corporation (NERC) as the capability to meet end-users electricity demand with a reasonable assurance of continuity and quality [1]. Further, NERC subdivides the reliability of power system into adequacy and security. Specifically, power system adequacy refers to "having sufficient resources to provide customers with a continuous supply of electricity at the proper voltage and frequency, virtually all of the time" [1]. Here, the "resources" include a combination of generation, transmission and distribution facilities, as well as demand side management (DSM) which may reduce end-users' electricity demand as needed. Power system security relates to the ability of bulk power systems to withstand unexpected disturbances, such as transmission line outage, loss of generator and transformer failure.

During the last decade, wind power has been the fastest in growth among all renewable energy resources [2]. With a prospective high penetration level, wind generation integration is expected to change dramatically the existing operating practices (e.g., unit commitment, economic dispatch and ancillary services procurement) that are critical to the adequacy of bulk power systems. Compared to conventional generation (e.g., thermal, hydro, nuclear), wind generation has three distinct characteristics: non-dispatchability, variability and uncertainty. Wind generation is generally nondispatchable, in the sense that the power output of a wind farm cannot be simply dispatched at the request of power grid operators. This is because, unlike conventional generators, the "fuel" of wind turbines, i.e., wind, cannot be controlled or stored, which is the largest challenges of wind power as pointed out in [3]. Due to the aforementioned characteristics, timely and accurate wind generation forecasting is critical to ensure that adequate resources for dispatch, ancillary services and ramping requirements are available all the time. Several balancing authorities in North America have been implementing wind power forecasting systems [4].

1.3   Wind Power Forecasting

To efficiently integrate wind power into power systems, wind power forecasting systems for different time scales (from minutes to weeks) need to be developed, depending on the system operations. For example, short-term wind power forecast is critical for optimization of the scheduling of conventional power plants by functions such as economic dispatch and unit commitment, and wind power forecast for longer time scales would be useful for the maintenance planning of large power plant components, wind turbines or transmission lines. The multi-scale uncertainty and variability of wind energy has made wind power forecast quite challenging [5–10]. Recent reports indicate that the state-of-the-art technology for wind power forecast often has an error of 15–20 % [4], which is a significant technical barrier for high penetration of renewable generation into the bulk power system.

State-of-the-art wind power forecast approaches include numerical weather prediction (NWP) models [3, 11], time-series models (e.g., autoregressive models [12] and Kalman filtering [13]), Markov chains [14, 15], and data mining [16, 17]. NWP models are currently used for long-term wind power forecast, as the computational complexity often renders NWP models intractable for short-term wind power forecast. Time-series models and data mining based regression models, while being able to provide continuous-value wind/solar power forecast, could suffer from high computational complexity. In addition, as shown in [18], large wind energy plants often experience changes in wind power output of 20 % of rated capacity over 1 h, namely wind ramps, and wind power forecast errors can be high in the presence of wind ramp events [19]. Bossavy et al. [20] investigate the skill of NWP ensembles to make probabilistic forecasts of ramp occurrence. Zheng et al. [21] propose a data mining approach to predict wind farm power ramp rates, while Zareipour et al. [22] propose a support vector machine (SVM) approach to predict the class of wind power ramp events.

Compared to other forecast models, finite-state Markov models strike a good balance between complexity and model accuracy. In particular, the transition probability matrix of Markov chains, which is used to provide distributional forecasts and point forecasts, can be gleaned from historical data [14]; when new data points are available online, it is also easy to update the transition probability matrix. As shown in recent studies [23–25], stochastic scheduling of power systems based on distributional forecasts can improve the system efficiency, in terms of reducing system reserves. To this end, distributional forecasts based on the Markov-chain-based forecast model are developed in [26]. In a nutshell, the statistical characteristics of renewable generation can be gleaned from historical data.

As the enabler of the energy economy, the smart grid with large-scale renewable generation integration will form the backbone of the next industrial revolution for sustainable growth. Making a paradigm shift, the proposed work takes a data analytics based approach to investigate spatio-temporal correlation structure and develops a stochastic modeling and optimization framework to evaluate both static and dynamic aspects of system reliability and security.

1.4 Organization of the Monograph

The rest of this monograph is organized as follows:

Chapters 2 and 3 focus on the development of short-term wind power forecast using a spatio-temporal analysis approach.

In Chap. 2, short-term forecast of wind farm generation is investigated by applying spatio-temporal analysis to extensive measurement data collected from a large wind farm, where multiple classes of wind turbines are installed. Specifically, using the data of the wind turbines' power outputs recorded across two consecutive years, graph-learning based spatio-temporal analysis is carried out to characterize the statistical distribution and quantify the level crossing rate of the wind farm's aggregate power output. Built on these characterizations, finite-state Markov chains are constructed for each epoch of 3 h and for each individual month, which accounts for the diurnal non-stationarity and the seasonality of wind farm generation. Short-term distributional forecasts and a point forecast are then derived by using the Markov chains and ramp trend information. The distributional forecast can be utilized to study stochastic unit commitment and economic dispatch problems via a Markovian approach. The developed Markov-chain-based distributional forecasts are compared with existing approaches based on high-order autoregressive models and Markov chains by uniform quantization, and the devised point forecasts are compared with persistence forecasts and high-order autoregressive model-based point forecasts.

In Chap. 3, the finite-state Markov models developed in Chap. 2 is enhanced by incorporating the wind ramp forecast. As wind ramps introduce significant uncertainty in wind power generation, reliable system operation requires accurate detection and forecast of wind ramps, especially at high wind generation penetration levels. A support vector machine (SVM) enhanced Markov model for short-term wind power forecast is developed, taking into account not only wind ramps but also the diurnal non-stationarity and the seasonality of wind farm generation. Specifically, using the historical data of the wind turbine power outputs recorded at an actual wind farm, multiple finite-state Markov chains that take into account the diurnal non-stationarity and the seasonality of wind generation are first developed to model the "normal" fluctuations of wind generation. To deal with the wind ramp dynamics, an SVM is then employed, based on one key observation from the measurement data that wind ramps often occur with specific patterns. Next, the forecast by the SVM is integrated cohesively into the finite-state Markov chain. Based on the SVM enhanced Markov model, both (short-term) distributional forecasts and point forecasts are then derived.

Chapter 4 investigates stochastic optimization of economic dispatch (ED) and interruptible load management using the proposed short-term distributional forecast of wind farm generation. Based on the distributional forecast model, the joint optimization of ED and interruptible load management is cast as a stochastic optimization problem. Additionally, a robust ED is formulated using an uncertainty set constructed based on the proposed distributional forecast, aiming to minimize the system cost for worst cases. The proposed stochastic ED is compared with three other ED schemes, namely the robust ED, the deterministic ED using the persistence wind generation forecast model, and the genie-aided ED with perfect wind generation forecasts. Numerical studies, using the IEEE Reliability Test System – 1996 and realistic wind measurement data from an actual wind farm, demonstrate the significant benefits obtained by leveraging the Markov-chain-based distributional forecast and the interruptible load management.

Chapter 5 concludes this monograph and gives perspectives for future research.

References

1. NERC, "Reliability standards: Glossary of terms," [Online] Available: http://www.nerc.com/page.php?cid=2|20.
2. "Accommodating high levels of variable generation," NERC Special Report, [Online] Available: www.nerc.com/files/IVGTF_Report_041609.pdf.
3. G. Giebel, R. Brownsword, G. Kariniotakis, M. Denhard, and C. Draxl, *The State of the Art in Short-Term Prediction of Wind Power - A Literature Overview*. ANEMOS.plus, 2011. [Online] Available: http://www.anemos-plus.eu/images/pubs/deliverables/aplus.deliverable_d1.2.stp_sota_v1.1.pdf.
4. D. Lew, M. Milligan, G. Jordan, and R. Piwko, "The value of wind power forecasting," *NREL Conference Paper CP-5500-50814*, Apr. 2011.
5. J. Zhang, B.-M. Hodge, and F. Anthony, "Investigating the correlation between wind and solar power forecast errors in the Western Interconnection," in *ASME 7th International Conference on Energy Sustainability and the 11th Fuel Cell Science, Engineering, and Technology Conference*, (Minneapolis, MN), 2013.
6. C. Yang and L. Xie, "A novel ARX-based multi-scale spatio-temporal solar power forecast model," in *North American Power Symp.*, pp. 1–6, 2012.
7. V. Kostylev and A. Pavlovski, "Solar power forecasting performance—towards industry standards," in *1st International Workshop on the Integration of Solar Power into Power Systems*, (Aarhus, Denmark), 2011.
8. C. Monteiro, T. Santos, L. A. Fernandez-Jimenez, I. J. Ramirez-Rosado, and M. S. Terreros-Olarte, "Short-term power forecasting model for photovoltaic plants based on historical similarity," *Energies*, vol. 6, no. 5, pp. 2624–2643, 2013.
9. M. Brabec, E. Pelikan, P. Krč, K. Eben, and P. Musilek, "Statistical modeling of energy production by photovoltaic farms," *JNL Energy & Power Eng.*, vol. 5, pp. 785–793, Sep. 2011.
10. S. Pelland, G. Galanis, and G. Kallos, "Solar and photovoltaic forecasting through post-processing of the Global Environmental Multiscale numerical weather prediction model," *Progress in Photovoltaics: Research and Applications*, vol. 21, no. 3, pp. 284–296, 2013.
11. C. Monteiro, H. Keko, R. Bessa, V. Miranda, A. Botterud, J. Wang, and G. Conzelmann, "A quick guide to wind power forecating: state-of-the-art 2009." [Online] Available: http://www.dis.anl.gov/pubs/65614.pdf, 2009.
12. P. Pinson and H. Madsen, "Adaptive modelling and forecasting of offshore wind power fluctuations with markov-switching autoregressive models," *Journal of Forecasting*, vol. 31, no. 4, pp. 281–313, 2012.
13. F. Cassola and M. Burlando, "Wind speed and wind energy forecast through kalman filtering of numerical weather prediction model output," *Applied Energy*, vol. 99, pp. 154–166, 2012.
14. G. Papaefthymiou and B. Klockl, "Mcmc for wind power simulation," *IEEE Trans. on Energy Convers.*, vol. 23, pp. 234–240, Mar. 2008.
15. A. Carpinone, R. Langella, A. Testa, and M. Giorgio, "Very short-term probabilistic wind power forecasting based on Markov chain models," in *Probabilistic Methods Applied to Power Systems (PMAPS), 2010 IEEE 11th International Conference on*, pp. 107–112, June 2010.
16. S. Santoso, M. Negnevitsky, and N. Hatziargyriou, "Data mining and analysis techniques in wind power system applications: abridged," in *Power Engineering Society General Meeting*, pp. 1–3, 2006.
17. A. Kusiak, H. Zheng, and Z. Song, "Wind farm power prediction: a data-mining approach," *Wind Energy*, vol. 12, no. 3, pp. 275–293, 2009.
18. C. Potter, E. Grimit, and B. Nijssen, "Potential benefits of a dedicated probabilistic rapid ramp event forecast tool," in *Power Systems Conference and Exposition, 2009. PSCE '09. IEEE/PES*, pp. 1–5, 2009.
19. C. Ferreira, J. Gama, L. Matias, A. Botterud, and J. Wang, "A survey on wind power ramp forecasting." Argonne National Laboratory Technical Report, 2010, available at http://www.dis.anl.gov/pubs/69166.pdf.

20. A. Bossavy, R. Girard, and G. Kariniotakis, "Forecasting ramps of wind power production with numerical weather prediction ensembles," *Wind Energy*, vol. 16, no. 3, pp. 51–63, 2013.

21. H. Y. Zheng and A. Kusiak, "Prediction of wind farm power ramp rates: A data-mining approach," *ASME Journal of Solar Energy Engineering*, vol. 131, pp. 031011.1–031011.8, 2009.

22. H. Zareipour, D. Huang, and W. Rosehart, "Wind power ramp events classification and forecasting: A data mining approach," in *Power and Energy Society General Meeting, 2011 IEEE*, pp. 1–3, IEEE, 2011.

23. "All island grid study – work stream 4, analysis of impacts and benefits." [Online]. Available at: http://www.dcenr.gov.ie/Energy/North-South+Co-operation+in+the+Energy+Sector/All+Island+Electricity+Grid+Study.htm, Jan. 2008.

24. "WILMAR (Wind Power Integration in Liberalised Electricity Markets)." [Online]. Available at: http://www.wilmar.risoe.dk/index.htm.

25. A. Papavasiliou, S. S. Oren, and R. P. O'Neill, "Reserve requirements for wind power integration: A scenario-based stochastic programming framework," *IEEE Trans. Power Syst.*, vol. 26, pp. 2197–2206, Nov. 2011.

26. M. He, L. Yang, J. Zhang, and V. Vittal, "A spatio-temporal analysis approach for short-term wind-farm power generation forecast," *IEEE Trans. Power Syst.*, vol. 28, no. 4, pp. 1611–1622, 2014.

Chapter 2
A Spatio-Temporal Analysis Approach for Short-Term Forecast of Wind Farm Generation

In this chapter, short-term forecast of wind farm generation is investigated by applying spatio-temporal analysis to extensive measurement data collected from a large wind farm where multiple classes of wind turbines are installed. Specifically, using the data of the wind turbines' power outputs recorded across two consecutive years, graph-learning based spatio-temporal analysis is carried out to characterize the statistical distribution and quantify the level crossing rate of the wind farm's aggregate power output. Built on these characterizations, finite-state Markov chains are constructed for each epoch of 3 h and for each individual month, which accounts for the diurnal non-stationarity and the seasonality of wind farm generation. Short-term distributional forecasts and a point forecast are then derived by using the Markov chains and ramp trend information. The distributional forecast can be utilized to study stochastic unit commitment and economic dispatch problems via a Markovian approach. The developed Markov-chain-based distributional forecasts are compared with existing approaches based on high-order autoregressive models and Markov chains by uniform quantization, and the devised point forecasts are compared with persistence forecasts and high-order autoregressive model-based point forecasts.

2.1 Introduction

A critical aspect in meeting the renewable portfolio standard (RPS) adopted by many states in the U.S. includes the integration of renewable energy sources, such as wind and solar [4]. Given the fact that the power outputs of wind turbines are highly dependent on wind speed, the power generation of a wind farm varies across multiple timescales of power system planning and operations. With increasing penetration into bulk power systems, wind generation has posed significant challenges for reliable system operations, because of its high variability and non-dispatchability [5]. Specifically, one key complication arises in terms of committing and dispatching conventional generation resources, when the short-term forecast of wind farm generation is not accurate. Currently, wind generation forecast for an individual wind farm typically has an error of 15–20 % [13], in sharp contrast to the case of load

© The Author(s) 2014
L. Yang et al., *Spatio-Temporal Data Analytics for Wind Energy Integration*,
SpringerBriefs in Electrical and Computer Engineering, DOI 10.1007/978-3-319-12319-6_2

forecast. When the actual wind generation is above the forecasted value, i.e., more conventional generation capacity has been committed than needed, it could result in less efficient set points for thermal units. In some cases, wind generation may need to be curtailed [6]. On the flip side of the coin, when the actual wind generation is less than the forecasted value, costly ancillary services and fast acting reserves have to be called upon. Therefore, it is imperative to develop accurate forecast approaches for wind farm generation.

State-of-the-art short-term wind power forecast approaches include time-series models (e.g., autoregressive models [24], Kalman filtering [3]), Markov chains [2, 21], and data mining [11, 27]. A comprehensive literature review on wind power forecast can be found in [7] and [16]. Time-series models and data mining-based regression models, while being able to provide continuous-value wind power forecast, could suffer from high computational complexity. Compared to other forecast models, finite-state Markov chains strike a good balance between complexity and modeling accuracy. In particular, the transition probability matrix of Markov chains, which is used to provide distributional forecasts and point forecasts, can be learned from historical data (e.g., by using the maximum likelihood estimation technique [21]); when new data points are available online, it is also easy to update the transition probability matrix. It is worth noting that one of state-of-the-art forecasting approaches is to utilize empirical distributions and the rich statistical information extracted from historical data (see [20, 23] and the references therein). Generally, empirical distribution of wind power data is non-Gaussian [12]. In [22], the logit transform is carried out as preprocessing, so that such a bounded time series can be studied by using autoregressive models in a Gaussian framework. In this paper, finite-state Markov chains are utilized to model the bounded wind power time series with a general probability distribution. It is worth noting that finite-state Markov chains inherently have bounded support, and the stationary distribution of a Markov chain can be general. Despite the appealing features of Markov chains, there is no existing studies to systematically design the state space of Markov chains for wind power. The proposed approach in this chapter addresses this issue by developing a general spatio-temporal analysis framework.

In this chapter, Markov-chain-based stochastic models for wind farm generation are developed for different seasons and for different epochs of the day across the whole year. From these Markov-chain-based stochastic models, short-term distributional forecasts and point forecasts of wind farm generation are obtained. The information used for forecasts includes both historical data and real-time data (the present wind farm generation). With a forecasting lead time of 10 min (or larger), these Markov-chain-based forecasts could be utilized for a variety of power system operation functions.

One key observation of this study is the wind farm spatial dynamics, i.e., *the power outputs of wind turbines within the same wind farm can be quite different, even if the wind turbines are of the same class and physically located close to each other.* The disparity in the power outputs of wind turbines may be due to the wake effect of wind speed, diverse terrain conditions, or other environmental effects. Motivated by this observation, graph-learning based spatial analysis is carried out to quantify the statistical distribution of *wind farm generation*, with rigorous characterization of wind farm spatial dynamics. Then, time series analysis is applied to quantify the level crossing rate (LCR) of the wind farm's aggregate power output. Finite-state Markov chains are then constructed, with the state space and transition matrix designed to capture both the spatial and temporal dynamics of the wind farm's aggregate power output. Based on [17], the distributional forecasts and the point forecasts of wind farm generation are provided by using the Markov chains and ramp trend information.

In this chapter, another finding of independent interest is that the tail probability of wind farm's aggregate power output exhibits a "power-law" decay with an exponential cut-off, where the power-law part has a much heavier tail than the Gaussian distribution. This indicates that one cannot simply apply the central limit theorem (CLT) to characterize the aggregate power output, because of the strong correlation across the power outputs of wind turbines within a wind farm.

The main contents of this chapter are summarized below:

- A general spatio-temporal analysis framework is developed, in which the spatial and temporal dynamics of wind farm generation are characterized by analytically quantifying the statistical distribution and the LCR.
- Built on the results of spatio-temporal analysis, a systematic approach for designing the state space of the Markov chain is introduced.

The rest of the chapter is organized as follows. A few critical observations from the measurement data are first discussed in Sect. 2.2. Spatio-temporal analysis and the design of Markov chains are presented in Sect. 2.3. Sect. 2.4 discusses the proposed Markov-chain-based forecast approach and numerical examples. A summary of the chapter is provided in Sect. 2.5. The main notation used in the chapter is summarized in Table 2.1.

2.2   Available Data and Key Observations

In this chapter, spatio-temporal analysis is carried out for a large wind farm with a rated capacity of $P_{ag}^{\max}=300.5$ MW. There are $M = 2$ classes of wind turbines in this wind farm, with $N_1 = 53$ and $N_2 = 221$, respectively. The power curves of the two turbine classes are provided in Fig. 2.1. For each class C_m, a meteorological tower (MET) H_m is deployed and co-located with a wind turbine, denoted by r_m. The power outputs of all wind turbines and the wind speeds measured at all METs are recorded every 10 min for the years 2009 and 2010. From the measurement data, several key observations can be made as follows.

2.2.1   Spatial Dynamics of Wind Farm

A critical observation from the measurement data is that the power outputs of wind turbines within the wind farm can be quite different. Figure 2.2 illustrates the power

Table 2.1 Summary of the main notation

Notation	Definition
t	Time index of measurement data
m	Index of wind turbine class and the corresponding meteorological tower (MET)
M	Number of wind turbine classes within the wind farm
C_m	Wind turbine class m
N_t	Number of measurement data
N_m	Number of wind turbines in C_m
H_m	MET for C_m
r_m	Wind turbine co-located with H_m in C_m
$W_m(t)$	Wind speed measured at H_m
$P_i(t)$	Power output of wind turbine i
$U_m(\cdot)$	Power curve of C_m, which maps $W_m(t)$ to $P_i(t)$, $\forall i \in C_m$
$P_{ag,m}(t)$	Aggregate power output of C_m
$P_{ag}(t)$	Aggregate power output of the wind farm
$P_{ag}^{\max}$	Rated capacity of the wind farm
$\overline{m}$	Index of the reference MET
$d_m(i)$	"distance" from node i to the root of the minimal spanning tree of C_m
α_m	linear regression coefficient for the parent-child turbine pairs of C_m
β_m	Linear regression coefficient for $W_m(t)$ as an affine function of $W_{\overline{m}}(t)$
$G_{pw}(\cdot)$	"power curve" of the wind farm, which maps $W_{\overline{m}}(t)$ to $P_{ag}(t)$
Γ	Wind farm generation level
γ	Wind speed level
$f_X(\cdot)$	Probability density function (PDF) of X
$F_X(\cdot)$	Cumulative density function (CDF) of X
$L_X(\cdot)$	Level crossing rate (LCR) function of X
$\mathcal{N}$	Standard normal random variable
$W_{\overline{m}}^{\mathcal{N}}(t)$	Gaussian transformation of $W_{\overline{m}}(t)$
ϕ	Regression coefficient of the first-order autoregressive (AR(1)) model
$\varepsilon(t)$	White noise of the AR(1) model
σ_ε	ariance of $\varepsilon(t)$
$\mathcal{S}$	State space of Markov chain (MC)
N_s	Number of states in $\mathcal{S}$
S_k	State k in $\mathcal{S}$, $k \in \{1, \cdots, N_s\}$
τ_k	Average duration of state S_k
$P_{ag,k}$	Representative generation level of state S_k
Q	Transition matrix of Markov chain

Table 2.1 (continued)

Notation	Definition
n_{ij}	Number of transitions from S_i to S_j encountered in the measurement data
$\Pr(A)$	Probability of an event A
$\mathbb{E}[X\|Y]$	Conditional expectation of X given Y
arg min	Argument of the minimum

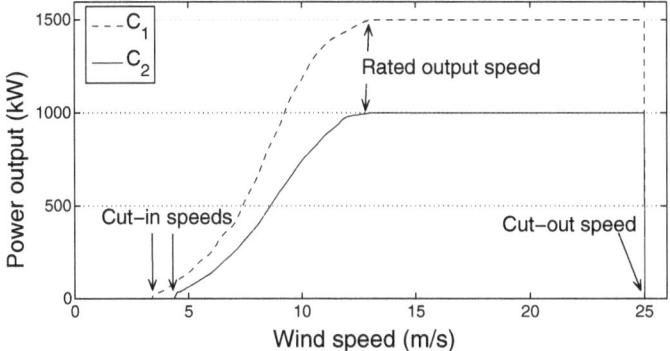

Fig. 2.1 Power curves for wind turbines from classes C_1 and C_2

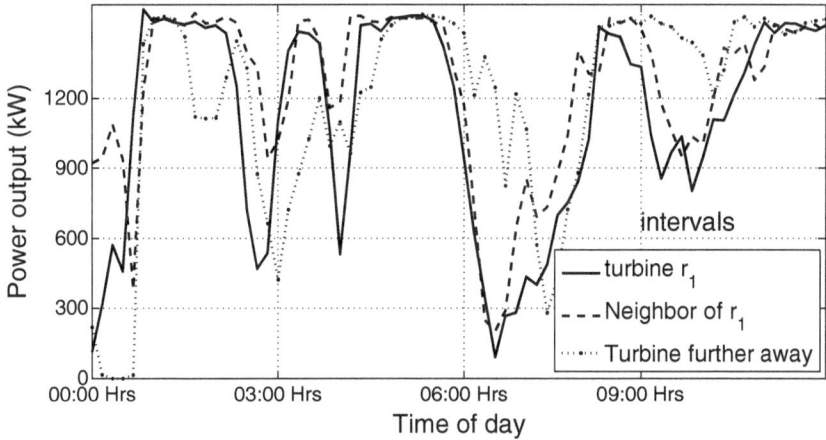

Fig. 2.2 Power outputs of three wind turbines in C_1

outputs of three wind turbines in C_1. It is clear that the power outputs are not equal, despite the geographic proximity of r_1 and its nearest neighbor (the disparity in the power outputs of the wind turbines belonging to C_2 has also been observed; the plots are not included for the sake of brevity). This disparity has been largely neglected in the existing literature.

Although the variable power outputs of wind turbines are not identical, it is reasonable to assume that they follow the same probability distribution if the wind

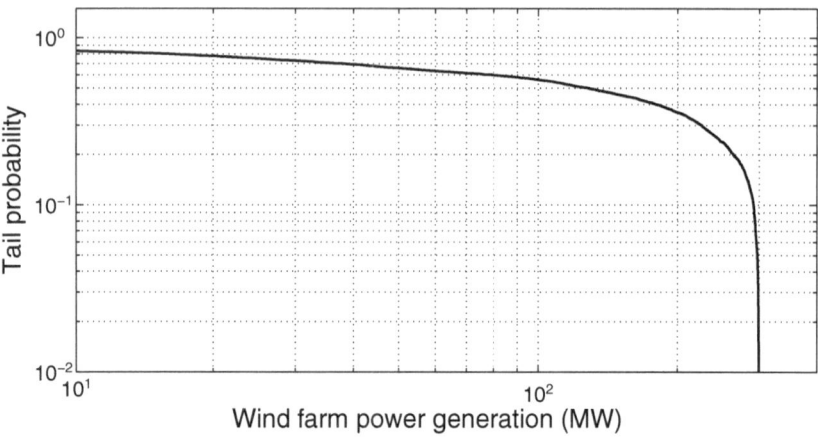

Fig. 2.3 Tail probability of the wind farm's aggregate power output

turbines are of the same class. A natural question here is whether the CLT, either the classic CLT or the generalized CLT, can be applied to characterize the probability distribution of the aggregate power output of a large number of wind turbines. To this end, the tail probability distribution of the wind farm's aggregate power output is examined and plotted in Fig. 2.3. As illustrated in Fig. 2.3, the tail probability demonstrates a "power-law" decay with an exponential cut-off and the power-law part has a much heavier tail than the Gaussian distribution. It is useful to note that this kind of tail behavior has been observed in many natural phenomena (e.g., size of forest fires) that have strong component-wise correlations [19]. Because of the strong correlation between the power outputs of wind turbines, particularly from adjacent wind turbines, the classic CLT cannot be applied to characterize the probability distribution of the wind farm's aggregate power output. In fact, even the "CLT under weak dependence" cannot be directly applied, despite the fact that the correlation between the power outputs of wind turbines weakens with the distance between them (the "mixing distance"). Hence, the probability distribution of the wind farm's aggregate power output cannot be characterized using the classic CLT; and it may not even be governed by stable laws [26]. With this insight, the proposed approach resorts to graphical learning methods to model the dependence structure in the power outputs of individual wind turbines and carries out spatio-temporal analysis accordingly.

2.2.2 Diurnal Non-Stationarity and Seasonality

Another key observation, as illustrated in Fig. 2.4, is the diurnal non-stationarity and the seasonality of wind farm generation. Specifically, it is observed that within each 3-h epoch, the probability distributions of wind farm generation over three consecutive 1-h intervals are consistent. However, these CDFs from different epochs of 3 h

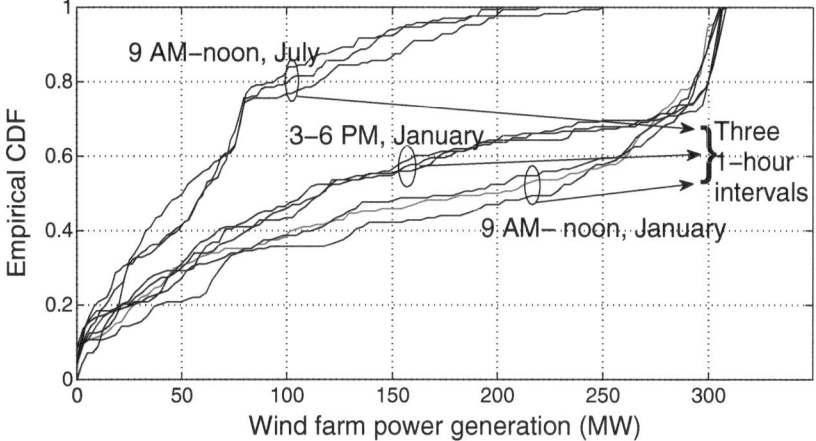

Fig. 2.4 Empirical distributions of wind farm generation over various 1-h intervals of different epochs of the day and different months

and different seasons can be quite different, indicating the non-stationarity of wind farm generation. Due to the non-stationary (empirical) distributions of wind farm generation, the distributional forecasts and point forecasts of wind farm generation, together with the developed models (Weibull distributions and Markov chains) used to derive distributional forecasts, can have quite different parameters for different months and different epochs. Therefore, it is necessary to develop forecast models *separately* for each month and each epoch (3 h for the wind farm considered here). Further, when estimating the parameters of Weibull distributions and Markov chains, relevant historical data, i.e., the historical data from the same month and the same epoch, can be used.

In what follows, data-driven analysis is carried out to characterize the spatial and temporal dynamics of the wind farm's aggregate power output. The data of the year 2009 is used in spatio-temporal analysis to guide the design of Markov chains, and the data of the year 2010 is used to assess the accuracy of the forecast provided by the proposed Markov-chain-based approach. Specifically, the 9 AM-noon epoch of January 2009 is used as an illustrative example in the following spatio-temporal analysis, since *this epoch exhibits the richest spatio-temporal dynamics, in the sense that the wind farm's aggregate power output during this epoch takes values ranging from 0 to the wind farm's rated capacity and exhibits the highest variability over time (quantified by LCR).*

2.2.3   Weibull Distribution of Wind Speed

In the existing literature, wind speed is usually characterized using Weibull distributions [28]. In this work, it is observed from the measurement data that the wind

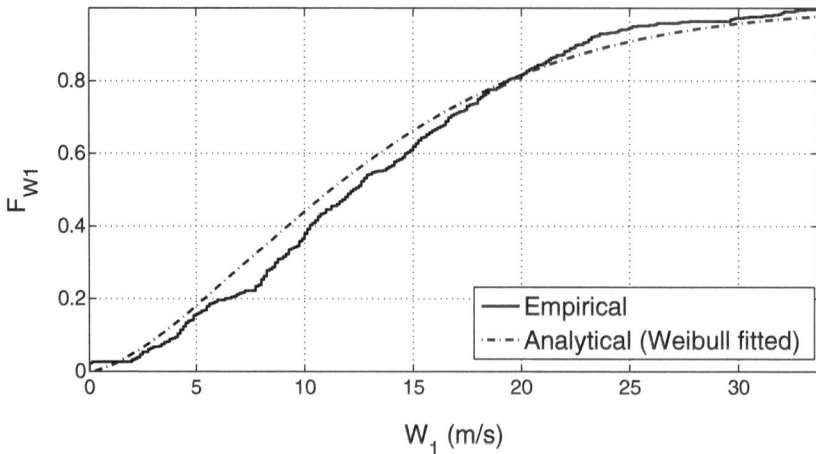

Fig. 2.5 Weibull-fitted CDF (λ=11.37, k=1.54) and empirical CDF of W_1 for the 9 AM-noon epoch of January 2009

speed W_m at each MET within the wind farm closely follows a Weibull distribution during each epoch, the probability density function (PDF) of which is given by:

$$f_{W_m}(x) = \frac{k}{\lambda} \left(\frac{x}{\lambda}\right)^{k-1} \exp^{-(x/\lambda)^k}, \quad \forall x \geq 0, \tag{2.1}$$

where k is the shape parameter and λ is the scale parameter. The fitted cumulative density function (CDF) and the empirical CDF of W_1 for the 9 AM-noon epoch of January 2009 are plotted in Fig. 2.5. The match between the empirical CDF and the fitted CDF suggests that the fitted Weibull distribution with the two parameters k and λ estimated from wind speed measurements can be utilized to analytical quantify wind speed dynamics. Under the developed spatio-temporal analysis framework, the fitted Weibull distributions of wind speed are also critical to the analytical characterizations of both the statistical distribution and the LCR of wind farm generation. The application of the fitted Weibull distributions of wind speed in the spatial analysis and the temporal analysis will be discussed in Sect. 2.3.1 and 2.3.2, respectively.

2.3 Spatio-Temporal Analysis of Wind Farm Generation

2.3.1 Spatial Analysis and Statistical Characterization

A key objective of spatial analysis is to characterize the statistical distribution of $P_{ag}(t)$. To this end, regression analysis is applied to the measurement data of each turbine's power output, so that $P_{ag}(t)$ could be expressed in terms of wind speed. Then, the analytical CDF of $P_{ag}(t)$ can be obtained from the fitted Weibull CDF of wind speed. In what follows, the key steps of spatial analysis are provided in detail.

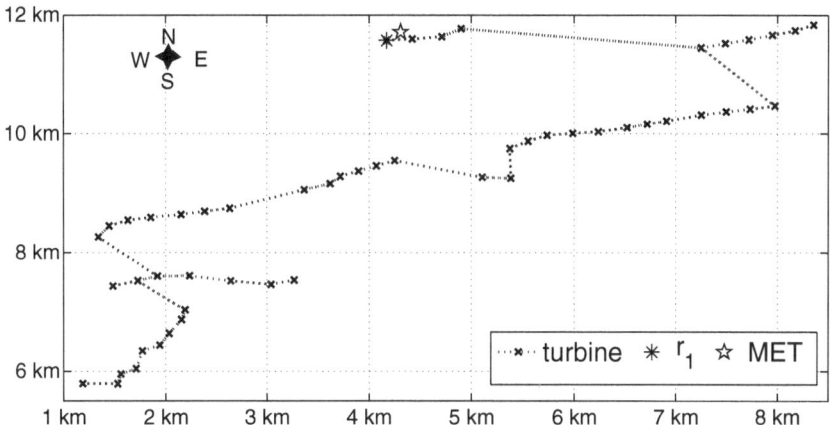

Fig. 2.6 MST of C_1 (with distance to the southwest corner of the wind farm)

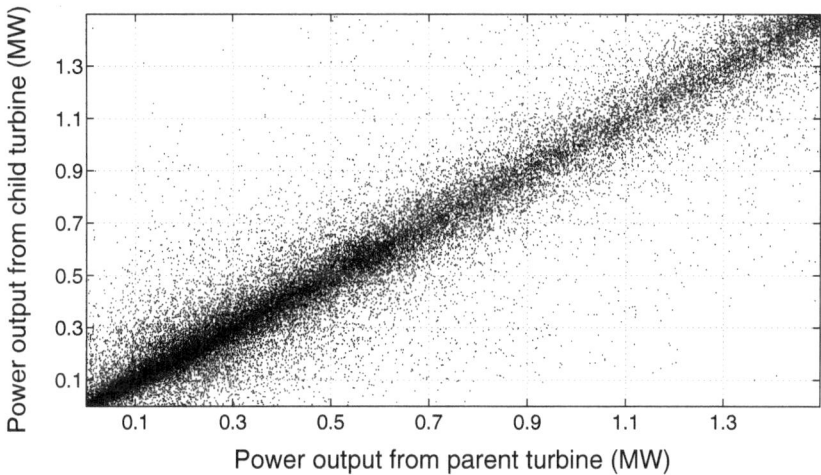

Fig. 2.7 Power outputs of parent-child turbine pairs of C_1 for the 9 AM-noon epoch of January 2009

Using the geographical information of wind turbine locations, a minimal spanning tree (MST) with r_m as the root node is constructed for each class C_m by using Prim's algorithm [25], as illustrated in Fig. 2.6. For each wind turbine i in C_m, there exists only one path from r_m to i in the MST of C_m. Define the node which is closest to i along this path as the 'parent' node of i. Another key observation from the measurement data is that an affine relationship exists between the power outputs of the parent-child turbine pairs for each class, with the case of C_1 illustrated in Fig. 2.7. Therefore, a coefficient α_m is introduced for C_m, and the linear regression model $P_k(t) = \alpha_m P_j(t)$ is used for each parent-child turbine pair (j,k) in C_m accordingly.

Further, define $d_m(i)$ as the number of the nodes (excluding node i) along the path from r_m to node i, then the linear regression model $P_i(t) = \alpha_m^{d_m(i)} P_{r_m}(t)$ can be used for any wind turbine i in C_m. The value of α_m is determined by applying the minimum mean square error (MMSE) principle to the aggregate power output of C_m, as follows:

$$\alpha_m = \arg\min_{\alpha} \frac{1}{N_t} \sum_t \left(P_{ag,m}(t) - \sum_{i \in C_m} \alpha^{d_m(i)} P_{r_m}(t) \right)^2. \qquad (2.2)$$

Similarly, an affine relationship between the wind speeds is also observed from the measurement data. For convenience, H_1 is chosen as the reference MET, i.e., $\overline{m} = 1$. Then, the linear regression models for wind speeds are given by $W_m(t) = \beta_m W_{\overline{m}}(t)$, where β_m is solved using the MMSE principle as follows:

$$\beta_m = \arg\min_{\beta} \frac{1}{N_t} \sum_t \left(W_m(t) - \beta W_{\overline{m}}(t) \right)^2. \qquad (2.3)$$

Using $P_{r_m}(t) = U_m\big(W_m(t)\big)$, the aggregate power output of the wind farm could be characterized as follows:

$$P_{ag}(t) = \sum_m P_{ag,m}(t) = \sum_m \sum_{i \in C_m} \alpha_m^{d_m(i)} U_m\big(\beta_m W_{\overline{m}}(t)\big) \triangleq G_{pw}\big(W_{\overline{m}}(t)\big). \qquad (2.4)$$

Due to the monotone characteristics of $U_m(\cdot)$, $G_{pw}(\cdot)$ is monotonically increasing. Therefore, the analytical CDF of $P_{ag}(t)$ can be obtained from the fitted Weibull distribution of $W_{\overline{m}}(t)$, given by $F_{P_{ag}}(\cdot) = F_{W_{\overline{m}}}\big(G_{pw}^{-1}(\cdot)\big)$. The analytical CDF and the empirical CDF of $P_{ag}(t)$ for the considered epoch are illustrated in Fig. 2.8.

It is worth noting that the linear regression models with homogeneous regression coefficients used here are motivated by the observation from the measurement data. The above regression analysis could be generalized by applying more general regression analysis methods. For example, each parent-child turbine pair can have a different linear regression coefficient or the parent-child turbine pairs can be analyzed by using different regression models.

2.3.2 Temporal Analysis and LCR Quantification

During each epoch, both the wind speed $W_{\overline{m}}(t)$ and the wind farm generation $P_{ag}(t)$ could be regarded as stationary stochastic processes. The LCR of a stochastic process is formally defined as the number of instances per unit time that the stochastic process crosses a level in only the positive/negative direction [29]. Intuitively, $L_{P_{ag}}(\cdot)$ quantifies how frequently $P_{ag}(t)$ transits between different generation levels. It will be apparent soon that $L_{P_{ag}}(\cdot)$, together with the statistical characterization $F_{P_{ag}}(\cdot)$, is critical in designing the state space representation of the Markov chains used for wind farm generation forecast.

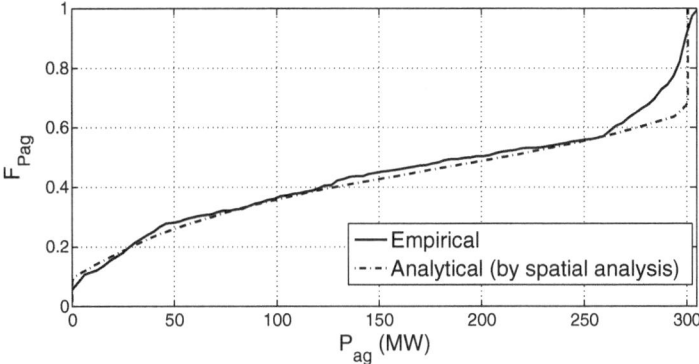

Fig. 2.8 CDF of $P_{ag}(t)$ for the 9 AM-noon epoch of January 2009

It is useful to note that due to the discontinuity in $F_{P_{ag}}(\cdot)$, as illustrated in Fig. 2.8, a smooth Gaussian transformation for $P_{ag}(t)$ is unattainable. Hence, the LCR of wind speed is first characterized. In order to quantify $L_{P_{ag}}(\cdot)$ analytically, $L_{W_{\overline{m}}}(\cdot)$ is first derived and converted to $L_{P_{ag}}(\cdot)$ by using the mapping defined in (2.4). To this end, autoregressive analysis is applied to $W_{\overline{m}}(t)$. As argued in [10], autoregressive analysis preceded by transforming the stationary non-Gaussian process $W_{\overline{m}}(t)$ to a Gaussian process can result in a better fit, compared with fitting to an autoregressive model directly. Therefore, $W_{\overline{m}}(t)$ is transformed to a standard normal random variable, given by

$$W_{\overline{m}}^{\mathcal{N}}(t) = F_{\mathcal{N}}^{-1}\left(F_{W_{\overline{m}}}(W_{\overline{m}}(t))\right), \tag{2.5}$$

A first-order autoregressive (AR(1)) model [1] is then fitted to $W_{\overline{m}}^{\mathcal{N}}(t)$:

$$W_{\overline{m}}^{\mathcal{N}}(t) = \phi W_{\overline{m}}^{\mathcal{N}}(t-1) + \varepsilon(t), \tag{2.6}$$

where the white noise term is modeled as a zero-mean Gaussian random variable $\varepsilon(t) \sim \mathcal{N}(0, \sigma_{\varepsilon}^2)$. It is worth noting that the above AR(1) model is not used for short-term wind speed prediction. Instead, it is used to quantify the LCR of wind speed. The parameters ϕ and σ_{ε} of the above AR(1) model can be estimated by solving the Yule-Walker equations [1]. Then, the LCR of $W_{\overline{m}}^{\mathcal{N}}(t)$ for a specific wind speed level γ $(\gamma > 0)$ can be calculated using the following steps:

$$L_{W_{\overline{m}}^{\mathcal{N}}}(\gamma) = \int_{-\infty}^{\gamma} \Pr\left(W_{\overline{m}}^{\mathcal{N}}(t) > \gamma \,|\, W_{\overline{m}}^{\mathcal{N}}(t-1) = w\right) f_{\mathcal{N}}(w) dw$$

$$= \int_{-\infty}^{\gamma} \Pr(\varepsilon(t) > \gamma - \phi w) f_{\mathcal{N}}(w) dw$$

$$= \int_{-\infty}^{\gamma} \left(1 - F_{\mathcal{N}}\left(\frac{\gamma - \phi w}{\sigma_{\varepsilon}}\right)\right) f_{\mathcal{N}}(w) dw. \tag{2.7}$$

Then, $L_{W_{\overline{m}}}(\cdot)$ can be obtained from $L_{W_{\overline{m}}^{\mathcal{N}}}(\cdot)$ using the inverse mapping of the strictly increasing function defined in (2.5). Further, using the monotonically increasing

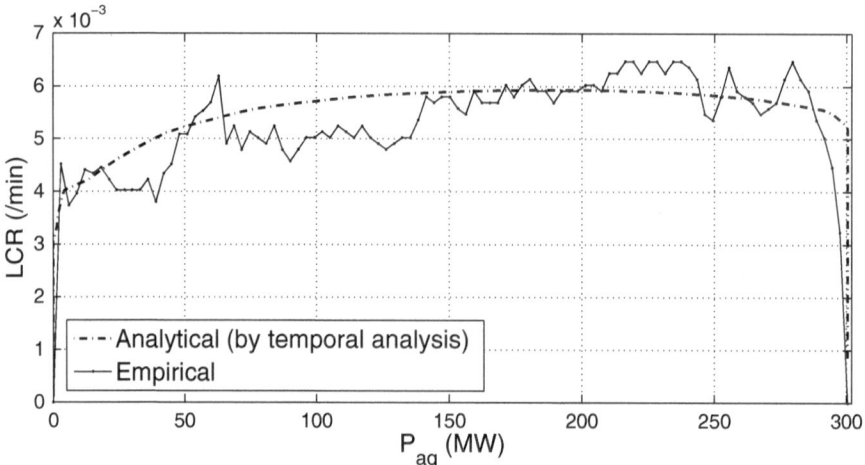

Fig. 2.9 LCR of $P_{ag}(t)$ for the 9 AM-noon epoch of January 2009

function defined in (2.4), the LCR of $P_{ag}(t)$ for a specific wind farm generation level Γ ($\Gamma \in (0, P_{ag}^{\max}]$) is given by:

$$L_{P_{ag}(t)}(\Gamma) = L_{W_{\overline{m}}^{\mathcal{N}}}\left(F_{\mathcal{N}}^{-1}(F_{W_{\overline{m}}}(G_{pw}^{-1}(\Gamma)))\right). \tag{2.8}$$

The procedure presented above completes the characterization of the analytical LCR of $P_{ag}(t)$ for an arbitrary epoch. The analytical LCR and the empirical LCR of $P_{ag}(t)$ for the 9 AM-noon epoch of January 2009 are illustrated in Fig. 2.9.

2.3.3 Markov Chain Model for Spatio-Temporal Wind Power

A critical step in developing the Markov-chain-based forecast approach is to capture the statistical distribution and the temporal dynamics of $P_{ag}(t)$ during each epoch using a Markov chain with the following characteristics:

- The Markov chain has finite states. Specifically, state S_k $(k = 1, \cdots, N_s)$ corresponds to a specific range of generation levels $[\Gamma_k, \Gamma_{k+1})$, with $\Gamma_1 = 0$ and $\Gamma_{N_s+1} = P_{ag}^{\max}$.
- The Markov chain is discrete-time and of order 1.

The above characteristics are adopted to make the Markov chains practical for forecasting applictions, so that forecast is made based on the most recent 10-min data only.

The objective of the Markov chain design is to determine the generation levels Γ_k $(k = 1, \cdots, N_s + 1)$ that defines the states, the transition matrix Q, and the representative generation level $P_{ag,k}$ for each state k. The procedure developed in [29] is utilized to design the state space. First, define τ_k as the average duration for which $P_{ag}(t)$ stays in S_k, given by:

$$\tau_k = \frac{F_{P_{ag}}\left(\Gamma_{k+1}\right) - F_{P_{ag}}\left(\Gamma_k\right)}{L_{P_{ag}}\left(\Gamma_{k+1}\right) + L_{P_{ag}}\left(\Gamma_k\right)}, \tag{2.9}$$

where $F_{P_{ag}}(\cdot)$ is the analytical CDF of $P_{ag}(t)$ that was characterized in spatial analysis, and $L_{P_{ag}}(\cdot)$ is the analytical LCR of $P_{ag}(t)$ derived in temporal analysis. Note that τ_k plays a critical role in the Markov chain model and determines how well the stochastic process $P_{ag}(t)$ is captured:

- A smaller value of τ_k suggests that $P_{ag}(t)$ is more likely to switch out of the state S_k within a 10-min slot, i.e., non-adjacent transitions are more likely to occur, *and hence the transitional behaviors of $P_{ag}(t)$ are not sufficiently captured by the discrete-time Markov chain.*
- If the values of τ_k $(k = 1, \cdots, N_s)$ are too large, there would be fewer states, indicating that the quantization by the Markov chain is too crude, and the corresponding forecast would be less accurate.

Therefore, a key objective of state space design is to make each of τ_k $(k = 1, \cdots, N_s)$ fall into a reasonable range [29]. However, it is challenging to achieve this design goal, especially when the closed-form expressions of $F_{P_{ag}}(\cdot)$ and $L_{P_{ag}}(\cdot)$ are unattainable. A practical solution adopted here is to introduce a constant τ and find the $N_s - 1$ variables $\{\Gamma_2, \Gamma_3, \cdots, \Gamma_{N_s}\}$ by solving (2.9) numerically with $\tau_k = \tau$, $\forall k \in \{1, \cdots, N_s - 1\}$. Once the state space $\mathcal{S}$ is designed, the transition probabilities can be estimated following the approach proposed in [21]. Specifically, the probability of a transition from S_i to S_j is given by

$$Q_{i,j} = \frac{n_{ij}}{\sum_{k=1}^{N_s} n_{ik}}, \quad i, j \in \{1, \cdots, N_s\}, \tag{2.10}$$

The representative generation level for each state S_k, $k \in \{1, \cdots, N_s\}$, is determined using the MMSE principle, given by (the time index of $P_{ag}(t)$ is dropped for simplicity):

$$P_{ag,k} = \underset{P_k}{\arg\min} \, \mathbb{E}\left[\left(P_k - P_{ag}\right)^2 | P_{ag} \in \left[\Gamma_k, \Gamma_{k+1}\right]\right], \tag{2.11}$$

Then, the representative generation level is given by:

$$P_{ag,k} = \frac{\int_{\Gamma_k}^{\Gamma_{k+1}} x f_{P_{ag}}(x) \mathrm{d}x}{F_{P_{ag}}\left(\Gamma_{k+1}\right) - F_{P_{ag}}\left(\Gamma_k\right)}. \tag{2.12}$$

The above procedure is applied to the 9 AM-noon epoch of January 2009, by choosing $\tau = 2$ min. The boundaries for each state are illustrated in Fig. 2.10a, and the corresponding transition probabilities are plotted in Fig. 2.11a. In [2, 21], the Markov chain for wind power (not in the context of *wind farm* generation) is obtained by uniform quantization. By choosing $\Gamma_{k+1} = P_{ag}^{\max} k / N_s$, $\forall k \in \{1, \cdots, N_s - 1\}$, the resultant state space, denoted by $\mathcal{S}_{\text{unif}}$, is compared with $\mathcal{S}$. From Fig. 2.10b, it is clear that higher values of τ_k are achieved for most of the states in $\mathcal{S}$. Hence, fewer non-adjacent transitions are incurred by $\mathcal{S}$, as can be seen from Fig. 2.11.

2.4   Markov-Chain-based Short-Term Forecast of Wind Farm Generation

As illustrated in Figs. 2.12 and 2.13, the proposed approach for short-term wind farm generation forecasting consists of two major steps: offline spatio-temporal analysis and online forecasting. These two steps utilize two types of information to provide both distributional forecasts and points forecasts. Specifically, in offline spatio-temporal analysis, the procedures presented in Sect. 2.3 are carried out on historical data of turbines' power output and wind speed, for each epoch and each month, to build multiple Markov chains by capturing the statistical characteristics from the historical data. It is worth noting that Weibull parameter estimation is part of spatio-temporal analysis. The inputs to the spatial analysis sub-step are the wind farm's geographical information and historical data of each wind turbine's power output. Historical data of wind turbines' power output and wind speed is used by the temporal analysis sub-step. In online forecasting, the Markov chains obtained are applied to the real-time measurement of wind farm generation to provide both distributional forecasts and point forecasts. Specifically, the transition probabilities of Markov chains determine the *conditional* probability distribution of future wind power $\hat{P}_{ag}(t+1)$, i.e., the probability distribution of $\hat{P}_{ag}(t+1)$ conditioned on the real-time wind power measurement $P_{ag}(t)$.

In what follows, short-term distribution forecasts and point forecasts are first derived by using the three inputs to the online forecasting step: (1) the Markov chain developed for the present epoch and month, (2) the wind farm's present aggregate power output $P_{ag}(t)$ at time t, and (3) short-term complementary information that can be utilized to enhance forecasting (e.g., ramp trend information). Then, the developed forecasting methods, with the parameters of the Markov chain models computed by using 2009 measurement data, are tested on the corresponding 2010 measurement data. For example, the forecasting method with the Markov chain developed based on the measurement data in the 9 AM-noon epochs of January 2009 will be applied to the measurement data in the 9 AM-noon epochs of January 2010 only.

2.4.1   Short-Term Distributional Forecasts and Point Forecasts

To derive a short-term forecast by using the Markov chain, it is worth noting that some complementary information can be utilized. One such complementary information is the ramp trend of wind farm generation. It is observed from available data that wind farm generation usually increases or decreases for several consecutive time-slots. Therefore, the ramp trend can be used to "steer" the transition of the Markov chain.

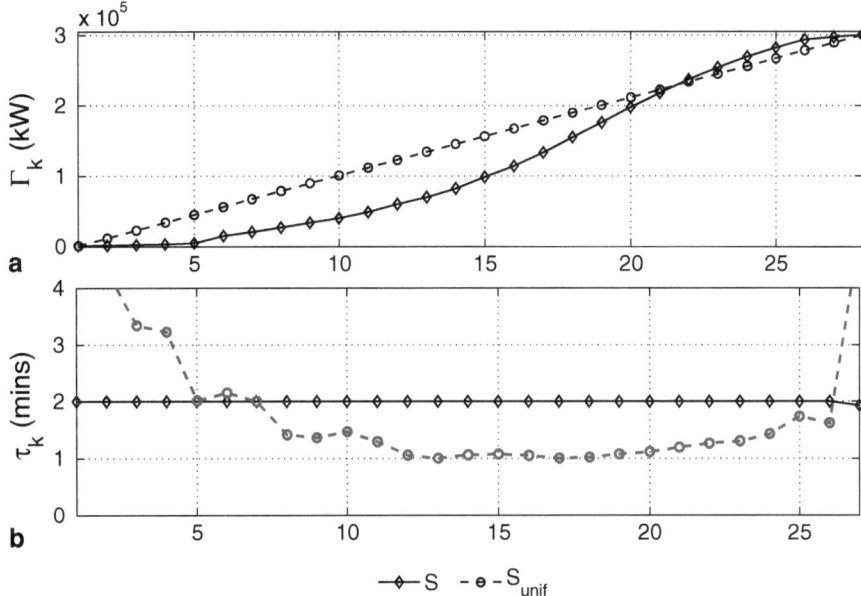

Fig. 2.10 Boundaries and average duration for each state of the Markov chain for the 9 AM-noon epoch of January 2009

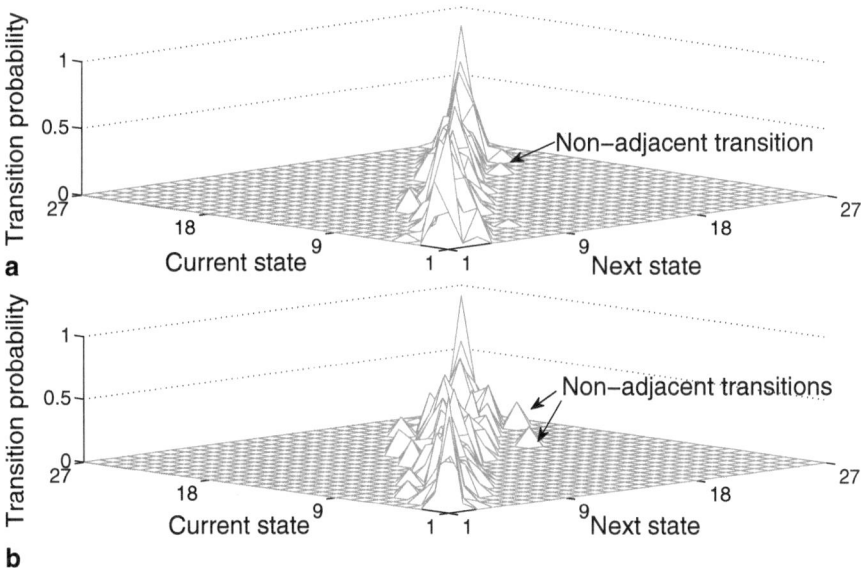

Fig. 2.11 Transition matrix **a** by spatio-temporal analysis **b** by uniform quantization, for the 9 AM-noon epoch of January 2009

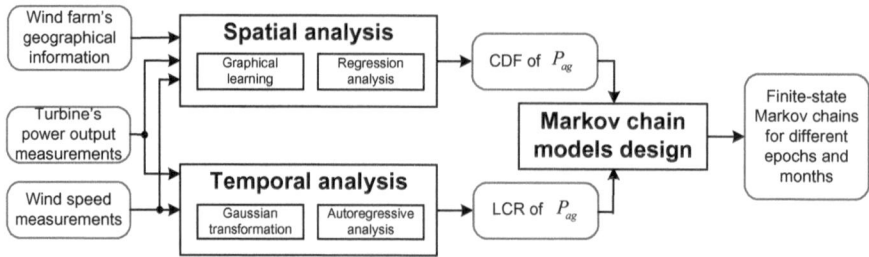

Fig. 2.12 Offline spatio-temporal analysis (carried out for each epoch and each month by using historical measurement data)

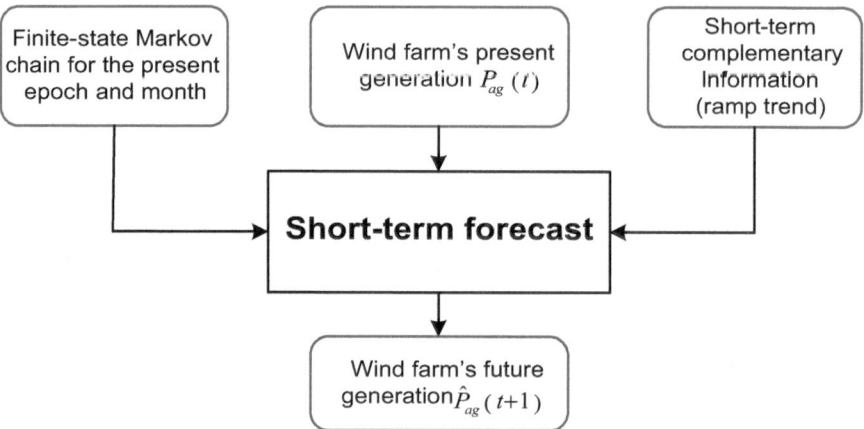

Fig. 2.13 Online short-term forecasting

2.4.1.1 Distributional Forecasts

Given the current 10-min wind farm generation data $P_{ag}(t)$, the state of the Markov chain at time t, denoted by $S(t)$, is determined by searching for a state k_0 so that $P_{ag}(t) \in [\Gamma_{k_0}, \Gamma_{k_0+1})$. Thus, $S(t+1)$ and hence $P_{ag}(t+1) = P_{ag,S(t+1)}$ are random variables that depend on the transition matrix Q, $S(t)$ and $R(t)$. Further, let $R(t) = -1$ denote a decreasing trend, and $R(t) = 1$ for the non-decreasing case. Then, the distributional forecast is given by

$$\Pr\big(P_{ag}(t+1) = P_{ag,j}\,|\,S(t), R(t)\big) = \begin{cases} \dfrac{Q_{k_0,j}}{\sum\limits_{k \geq k_0}^{N_s} Q_{k_0,k}}, & \text{if } R(t) = 1 \text{ and } j \geq k_0 \\[4mm] \dfrac{Q_{k_0,j}}{\sum\limits_{k=1}^{k_0-1} Q_{k_0,k}}, & \text{if } R(t) = -1 \text{ and } j < k_0 \\[4mm] 0, & \text{otherwise.} \end{cases}$$

$$(2.13)$$

2.4.1.2   Point Forecasts

From the above distributional forecast, a point forecast can be derived by using the MMSE principle:

$$\hat{P}_{ag}(t+1) = \underset{P_{ag}}{\text{argmin}}\ \mathbb{E}\left[\left(P_{ag} - P_{ag,S(t+1)}\right)^2 | S(t), R(t)\right] \tag{2.14}$$

Then, the solution to the above problem is given by:

$$\hat{P}_{ag}(t+1) = \begin{cases} \dfrac{\sum\limits_{k \geq k_0}^{N_s} P_{ag,k} Q_{k_0,k}}{\sum\limits_{k \geq k_0}^{N_s} Q_{k_0,k}}, & \text{if } R(t) = 1 \\[4mm] \dfrac{\sum\limits_{k=1}^{k_0-1} P_{ag,k} Q_{k_0,k}}{\sum\limits_{k=1}^{k_0-1} Q_{k_0,k}}, & \text{if } R(t) = -1 \end{cases} \tag{2.15}$$

which is exactly the mean value of the Markov chain conditioned on the currently observed state and the ramp trend.

2.4.2   Numerical Examples

2.4.2.1   Distributional Forecasts

The continuous rank probability score (CRPS) is utilized to quantitatively assess the performance of Markov-chain-based distributional forecasts, given by:

$$\text{CRPS} = \frac{1}{N_t} \sum_t \int_0^{P_{ag}^{\max}} \left(\hat{F}(x) - H\left(x - P_{ag}(t)\right)\right)^2 dx, \tag{2.16}$$

where N_t is the total number of data points, $\hat{F}(x)$ is the CDF obtained by using the Markov-chain-based distributional forecast, and $H(x - P_{ag}(t))$ is the unit step function, which takes value 0 when $x < P_{ag}(t)$ and takes value 1 when $x \geq P_{ag}(t)$. Basically, *a higher CRPS value suggests that the distributional forecast is less accurate*. By using the above definition, the CRPS value of the Markov-chain-based distributional forecast over all the 52560 ($365 \times 24 \times 6$) data points of the year 2010 is calculated. The CRPS of the Markov-chain-based distributional forecast over the data points of the year 2010 is provided in Table. 2.2. Since one main objective of this work is to develop Markov-chain-based distributional forecasting models, the Markov chain developed by the existing approach [21, 2] (uniform quantization) is used as a benchmark. The Markov chain developed by the proposed spatio-temporal analysis with the design parameter $\tau = 2$ (column 3 of Table. 2.2) has a CRPS that is

Table 2.2 CRPS of distributional forecasts over the testing data points of the year 2010

	MC (unif.)	MC ($\tau = 2$)	MC ($\tau = 1$)	AR (Gaussian)	AR (Log-normal)
CRPS	7.14 MW	6.27 MW	6.09 MW	6.89 MW	6.54 MW

13 % less than that of the benchmark Markov chain that has the same number of states designed by uniform quantization (column 2 of Table. 2.2). By reducing the design parameter τ to 1, the forecasting performance of the Markov chain developed by the proposed spatio-temporal analysis (column 4 of Table. 2.2) is further improved.

The proposed Markov-chain-based distributional forecasts are also compared with the distributional forecasts based on high-order AR models. Here, two high-order AR models with a truncated Gaussian distribution and a truncated log-normal distribution are considered. The high-order AR model with a Gaussian distribution is adopted from [24] by considering one regime, and then the support of the Gaussian distribution is truncated into $\left[0, P_{ag}^{\max}\right]$. The procedure for building AR models with truncated log-normal distributions can be found in [22]. Specifically, the order of the AR models are determined by using the partial autocorrelation functions of the wind power time series [8]. Then, the recursive least square algorithm [8] is applied to calculate the regressive coefficients, the predicted wind power $\hat{P}_{ag}(t)$ (the point forecast of the AR model), and the variance of innovation $\mathbf{C}$. Finally, by using $\hat{P}_{ag}(t)$ as the mean and $\mathbf{C}$ as the variance of a Gaussian distribution or a log-normal distribution which is truncated into $\left[0, P_{ag}^{\max}\right]$, the wind power distributional forecasts can be obtained. The CRPS values of the distributional forecasts based on high-order AR models are calculated by using (2.16), and are shown in Table 2.2. It can be seen from Table 2.2 that the Markov-chain-based distributional forecasts with the design parameter $\tau = 1$ (column 4 of Table 2.2) achieves a CRPS value that is 11.6 and 6.9 % lower than those of the AR-based distributional forecasts (column 5 and column 6 of Table 2.2), respectively. The reason for this improvement of Markov-chain-based distributional forecast is that the conditional probability distributions provided by Markov chains do not assume the shape of the distribution (and thus can be re- garded as "non-parametric" distributional forecasts in literature [23]). Therefore, by using the transition probability estimated from historical data, Markov chains can provide more accurate distributional forecasts than those based on assumed para- metric distributions (e.g., Gaussian, β and log-normal distributions). The superiority of non-parametric distributional forecasts over parametric ones is also discussed in [23] and references therein. In summary, the improvement of the developed Markov- chain-based approach over other approaches can be attributed to the rigorous design of Markov chains and transition probabilities, which in turn utilizes the analytical results from spatio-temporal analysis.

To further examine the performance of the developed Markov-chain-based distri- butional forecasting method over different epochs and different month, the median and percentiles of the CRPS values over the data points for each month or each epoch is computed. In the box plots of Figs. 2.14 and 2.15, the central bar in a box

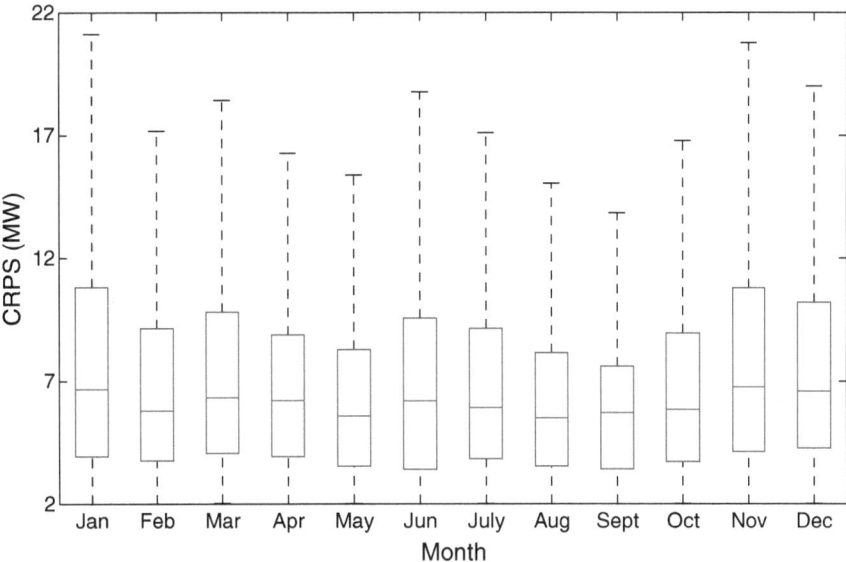

Fig. 2.14 Statistics of CRPS over all months of the year 2010

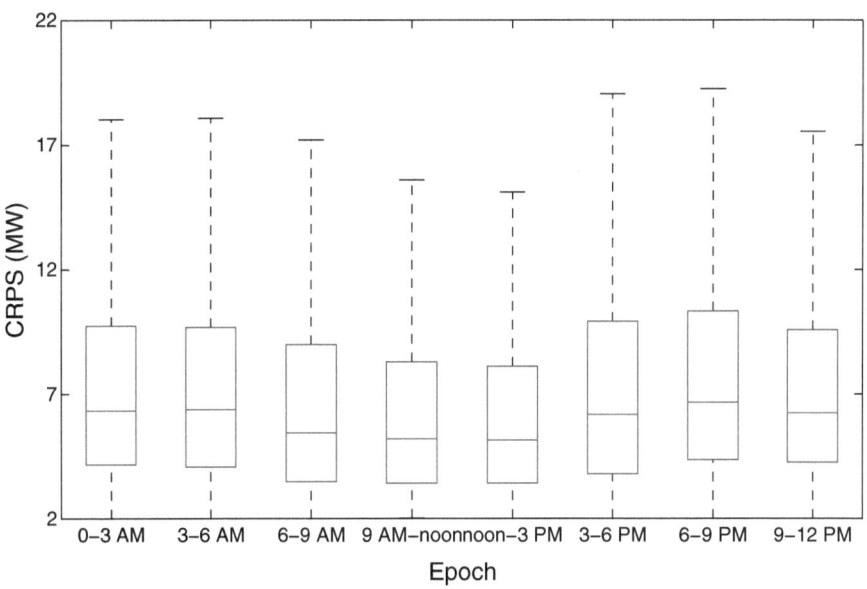

Fig. 2.15 Statistics of CRPS over all 8 epochs of the year 2010

represents the median value of the CRPS values over all data points that fall into a specific epoch or a specific month. The top edge and bottom edge of a box represent the 25th and 75th percentiles, respectively. The top bar and bottom bar correspond to

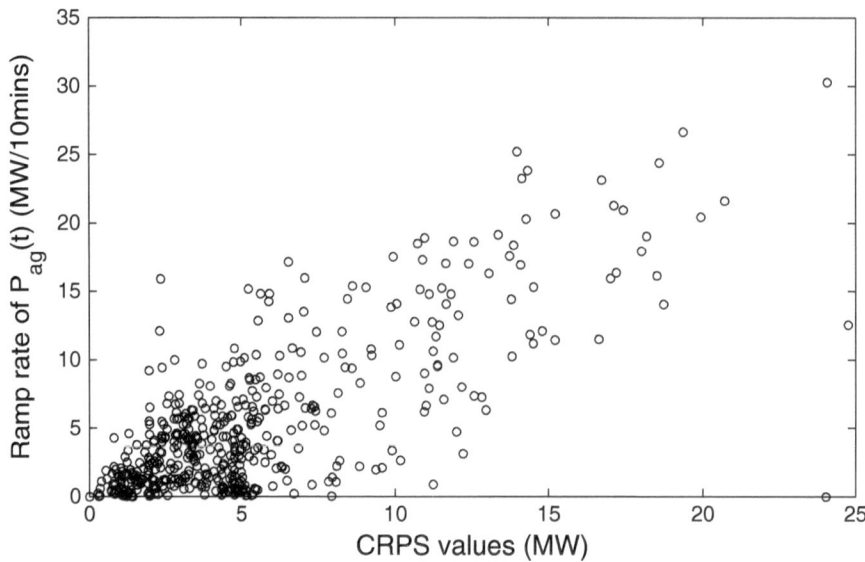

Fig. 2.16 Correlation between the ramp rates of $P_{ag}(t)$ and the CRPS values of distributional forecast

the extremes calculated from 1.5 interquartile ranges. It is observed from Fig. 2.15 that the medians and standard deviations of the CRPS values are a little higher during afternoon-night epochs. Figure 2.14 shows that the medians of the CRPS values have little variability across different months, and the standard deviations of the CRPS values are slightly higher across the winter season. Another key observation from the results of numerical experiments is that *the CRPS of the Markov-chain-based distributional forecast over a realized data points $P_{ag}(t)$ is highly dependent on the ramp rate of $P_{ag}(t)$ at time t*. Here, the ramp rate of $P_{ag}(t)$ is defined as the absolute value of the change in the wind farm generation in a 10-min slot. For example, the ramp rate of $P_{ag}(t)$ at time t is given by $|P_{ag}(t) - P_{ag}(t-1)|$. By using the data points of the year 2010, the corresponding pairs of ramp rates and CRPS values are plotted in Fig. 2.16. It is observed that the ramp rates of $P_{ag}(t)$ and the CRPS values of the Markov-chain-based distributional forecast over a realized data points $P_{ag}(t)$ follows a *positive correlation*. The above observation also explains the "phase transition" from the noon-3 PM epoch to the 3–6 PM in Fig. 2.15, i.e., the increased wind ramp caused by the sudden change in diurnal heating/vertical mixing conditions [18]. In summary, the statistics (especially the median value) of the CRPS values vary slightly differently over different months and epochs, which suggests that the developed Markov-chain-based distributional forecasting methods deliver consistent forecasting performance across the entire year.

Further, three episodes of prediction intervals are plotted to better illustrate the developed Markov-chain-based distributional forecasts. According to the above observation, three representative time periods are chosen: (1) the 0–3 AM epoch of

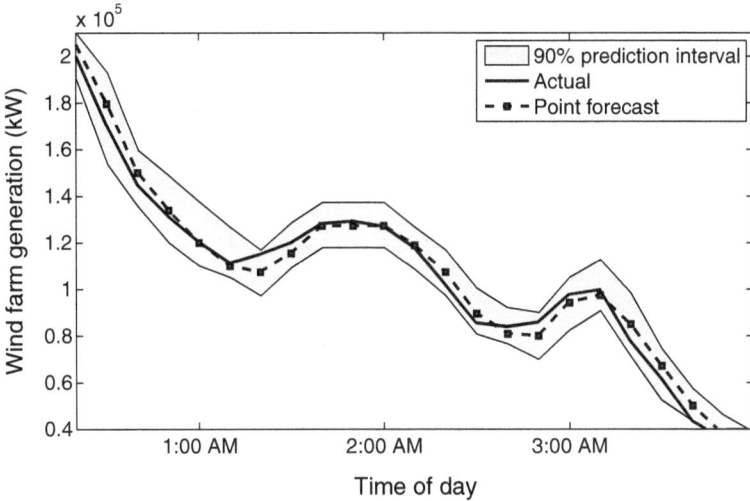

Fig. 2.17 Ten minutes distributional forecasts on January 23rd, 2010

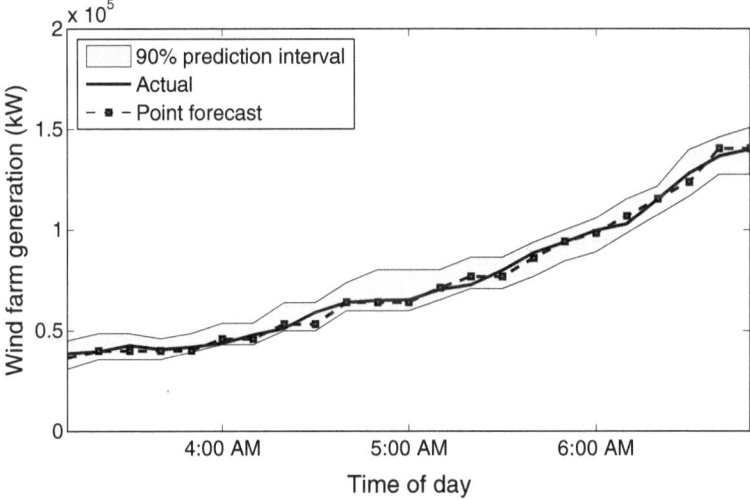

Fig. 2.18 Ten minutes distributional forecasts on January 16th, 2010

January 23rd, (2) the 3–6 PM epoch of January 16th, and (3) the 3–6 PM epoch of April 16th. The first period in January 23rd is chosen since it has much higher average ramp rate than other January days, and the 0–3 AM epoch experienced a large down-ramp from 75 to 25 % of the rated capacity. The second period is chosen because January and the 3–6 PM epoch have the highest median CRPS value (i.e., least accurate forecasts), and the CRPS value of January 16th is mostly close to the corresponding median value. The third period is chosen due to similar reasons as

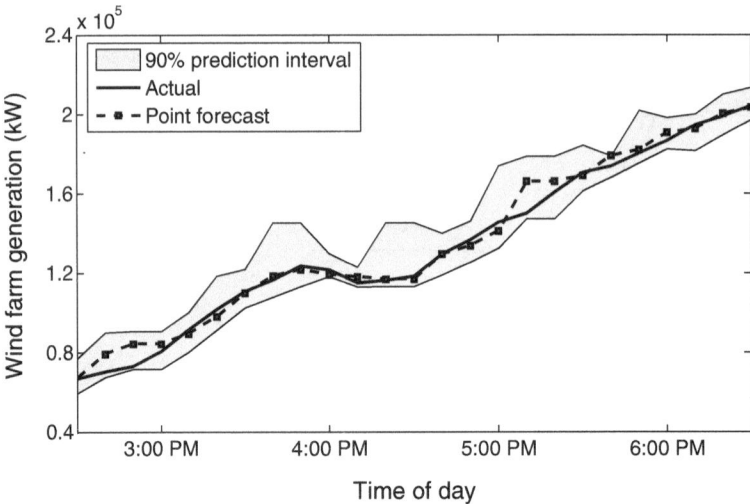

Fig. 2.19 Ten minutes distributional forecasts on April 16th, 2010

the second period, except that April is the month that has the least CRPS values. Figures 2.17, 2.18, and 2.19 illustrate the 90 % prediction intervals obtained by the developed Markov-chain-based distributional forecasts. It is observed at for all three representative periods, the realized wind farm generation reasonably lies in the 90 % prediction intervals.

2.4.2.2 Point Forecasts

By comparing the point forecast $\hat{P}_{ag}(t)$ with the actual wind farm generation $P_{ag}(t)$, forecast errors are quantified by mean absolute error (MAE), defined as

$$\text{MAE} = \frac{1}{N_t} \sum_t |P_{ag}(t) - \hat{P}_{ag}(t)|, \qquad (2.17)$$

mean absolute percentage error (MAPE), defined as

$$\text{MAPE} = \frac{\sum_t |P_{ag}(t) - \hat{P}_{ag}(t)|}{\sum_t P_{ag}(t)}, \qquad (2.18)$$

and root mean square error (RMSE), defined as

$$\text{RMSE} = \sqrt{\frac{\sum_t |P_{ag}(t) - \hat{P}_{ag}(t)|^2}{N_t}}. \qquad (2.19)$$

Besides AR models, two point forecast approaches are used as benchmark:

Table 2.3 Ten minutes point forecast error of wind farm generation (all test data of the year 2010 is used)

Error	Persistence	MC (unif)	MC ($\tau = 2$)	MC ($\tau = 1$)	AR
MAE	6.98 MW	7.14 MW	6.83 MW	6.62 MW	6.38 MW
MAPE	7.31 %	7.48 %	7.15 %	6.93 %	6.68 %
RMSE	11.18 MW	11.58 MW	10.89 MW	10.56 MW	10.25 MW

Table 2.4 Ten minutes point forecast error of wind farm generation over the period shown in Fig. 2.17

Error	Persistence	MC (unif)	MC ($\tau = 2$)	MC ($\tau = 1$)	AR
MAE	13.26 MW	13.83 MW	9.97 MW	9.59 MW	9.39 MW
MAPE	11.61 %	12.1 %	8.73 %	8.4 %	8.22 %
RMSE	15.81 MW	16.26 MW	12.81 MW	12.23 MW	11.94 MW

Table 2.5 Ten minutes point forecast error of wind farm generation over the period shown in Fig. 2.18

Error	Persistence	MC (unif)	MC ($\tau = 2$)	MC ($\tau = 1$)	AR
MAE	4.6 MW	4.71 MW	4.54 MW	4.32 MW	4.28 MW
MAPE	6.28 %	6.43 %	6.2 %	5.9 %	5.84 %
RMSE	6.16 MW	6.32 MW	6.09 MW	5.91 MW	5.86 MW

- persistence forecast [15]: $\hat{P}_{ag}(t + 1) = P_{ag}(t)$;
- forecast by Markov Chain with uniform quantization.

The proposed Markov-chain based forecast method is compared with several state-of-the-art approaches. Specifically, the wind power data used for forecast is first mapped to the state space designed by following the procedure in Sect. 2.3.3. Then, point forecasts are obtained by using the representative generation levels of corresponding states.

The test results by using the data for the year 2010 and the three selected epochs are provided in Tables 2.3, 2.4, 2.5, and 2.6, respectively. It is observed that the Markov chains based on uniform quantization give less accurate forecast than persistence forecast. This can be attributed to the uniform quantization not considering the spatio-temporal dynamics of wind farm generation. Also note that the proposed Markov-chain-based forecast approach has improved accuracy compared to the persistence forecast approach, and comparable accuracy to the AR-based approach. Further, the statistics of the absolute error of the Markov-chain-based point forecasts over different months and different epochs are illustrated in Fig. 2.20 and 2.21, respectively. It can be seen from Figs. 2.20 and 2.21 that the developed

Table 2.6 Ten minutes point forecast error of wind farm generation over the period shown in Fig. 2.19

Error	Persistence	MC (unif)	MC ($\tau = 2$)	MC ($\tau = 1$)	AR
MAE	6.02 MW	6.31 MW	4.95 MW	4.81 MW	4.73 MW
MAPE	4.64 %	4.86 %	3.82 %	3.71 %	3.65 %
RMSE	6.86 MW	7.17 MW	5.73 MW	5.41 MW	5.23 MW

Markov-chain-based point forecasting methods perform consistently across the entire year.

Another key observation from Tables 2.3 and 2.5 is that smaller values of τ leads to higher forecast accuracy of the Markov chains, at the cost of higher complexity of the Markov chains (in terms of the number of states). The trade-off between the forecast accuracy and the complexity of the Markov chain for the 9 AM-noon epoch of January 2010 is illustrated in Fig. 2.22.

From the results presented above, it can be seen that the proposed distributional forecast approach outperforms the high-order AR-based distributional forecasts with Gaussian and log-normal distributions. This is because the proposed spatio-temporal analysis extracts from historical data the rich statistical information of wind farm generation, and accordingly the corresponding Markov chain models can provide more accurate distributional forecasts than AR-based models with assumed Gaussian and log-normal distributions. Further, the proposed point forecasts have a lightly higher mean absolute error (MAE) than those of high-order AR-based forecasts. However, note that one main objective of this study is to develop Markov-chain-based distributional forecasts that can be used for economic dispatch in the presence of wind generation uncertainty [9, 14], in which a good balance is needed between computational complexity and modeling accuracy. Here, computational complexity involves both the computational effort for building and utilizing the forecasting models to provide distributional forecasts and the computational effort for solving stochastic economic dispatch problems by using these distributional forecasts. Therefore, compared to AR-based distributional forecast methods, the developed Markov chain models are more suitable for stochastic economic dispatch, because the computational burden of using continuous distributions of AR-based forecasts for stochastic economic dispatch would be significantly higher. Moreover, even though the computational effort of using AR-based distributional forecasts can be reduced by applying quantization (i.e., 0–300 MW quantized into 50–70 states for the cases in this study) and scenario reduction, the quantization error would cause the quantized AR-based forecasts to be even less accurate than the proposed Markov-chain-based forecasts. In summary, the proposed Markov-chain-based distributional forecast approach achieves higher accuracy than existing approaches, and the well-balanced complexity and accuracy of Markov chain models make them an ideal tool to study stochastic economic dispatch problems.

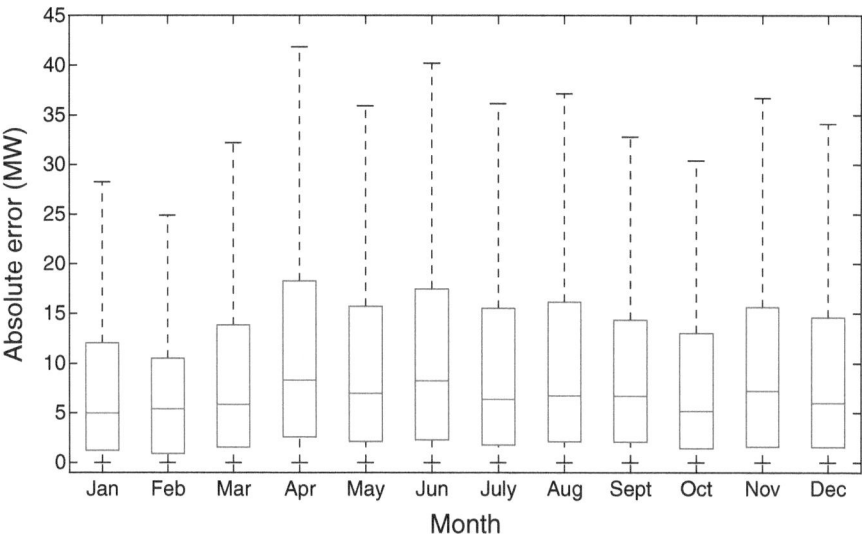

Fig. 2.20 Statistics of absolute error over all months of the year 2010

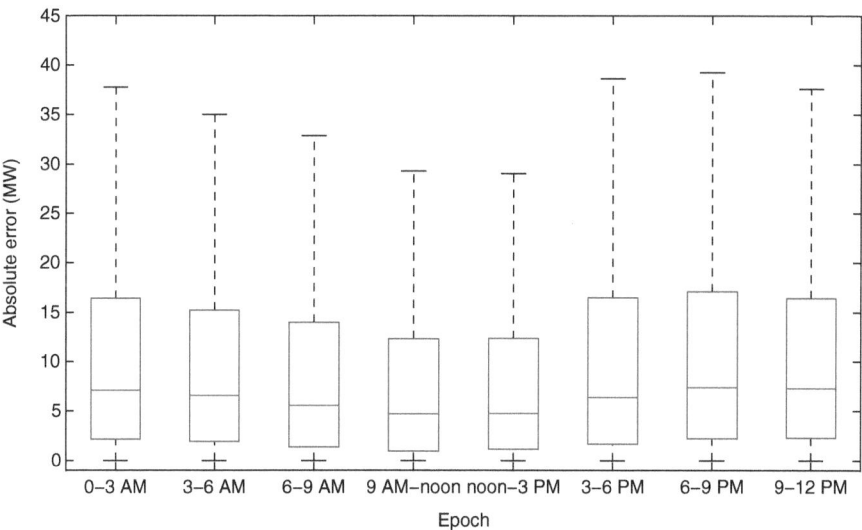

Fig. 2.21 Statistics of absolute error over all 8 epochs of the year 2010

2.5   Summary

In this chapter, a general spatio-temporal analysis framework is developed for wind
farm generation forecast, in which finite-state Markov chain models are derived.
The state space, transition matrix and representative generation levels of the Markov

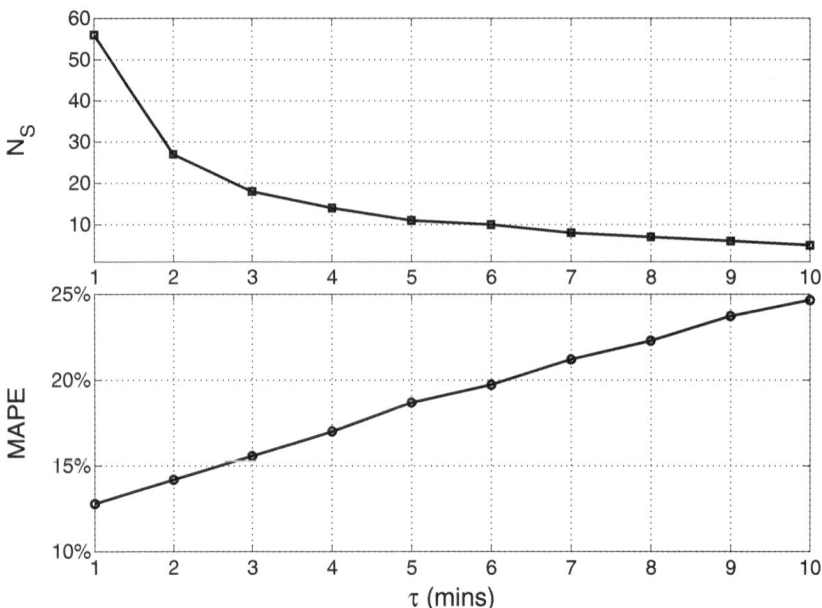

Fig. 2.22 Number of states and the forecast error of Markov chains at various τ for the January 9 AM-noon epoch)

chains are optimized by using a systematic approach. The short-term distributional forecast and point forecast are derived by using the Markov chains and the ramp trend information. The developed distributional forecast can be directly integrated into the problems of unit commitment and economic dispatch with uncertain wind generation, so that these problems can be studied in a general Markov-chain-based stochastic optimization framework.

In Chap. 4, we investigate power system economic dispatch with wind farm generation by utilizing a realistic test system and the Markov-chain-based distributional forecasts of wind farm generation. The distributional forecasts of wind farm generation are integrated into a stochastic programming framework of multi-period economic dispatch, so as to optimize the dispatch decisions over the operating horizon. The impact of the forecast errors of wind farm generation on economic dispatch is also studied.

References

1. G. E. P. Box and G. M. Jenkins, *Time Series Analysis: Forecasting and Control, 2nd ed.* San Francisco: Holden-Day, 1976.
2. A. Carpinone, R. Langella, A. Testa, and M. Giorgio, "Very short-term probabilistic wind power forecasting based on Markov chain models," in *Probabilistic Methods Applied to Power Systems (PMAPS), 2010 IEEE 11th International Conference on*, pp. 107–112, June 2010.
3. F. Cassola and M. Burlando, "Wind speed and wind energy forecast through Kalman filtering of numerical weather prediction model output," *Applied Energy*, vol. 99, pp. 154–166, 2012.
4. K. S. Cory and B. G. Swezey, "Renewable portfolio standards in the states: balancing goals and implementation strategies," *NREL Technical Report TP-670-41409*, Dec. 2007.
5. E. A. DeMeo, G. A. Jordan, C. Kalich, J. King, M. R. Milligan, C. Murley, B. Oakleaf, and M. J. Schuerger, "Accommodating wind's natural behavior," *IEEE Power Energy Mag.*, vol. 5, pp. 59–67, Nov.–Dec. 2007.
6. S. Fink, C. Mudd, K. Porter, and B. Morgenstern, "Wind energy curtailment case studies." NREL Subcontract Report SR-550-46716, Oct. 2009.
7. G. Giebel, R. Brownsword, G. Kariniotakis, M. Denhard, and C. Draxl, *The State of the Art in Short-Term Prediction of Wind Power—A Literature Overview.* ANEMOS.plus, 2011. [Online] Available: http://www.anemos-plus.eu/images/pubs/deliverables/aplus.deliverable_d1.2.stp_sota_v1.1.pdf.
8. M. H. Hayes, *Statistical Digital Signal Processing and Modeling.* Wiley, 1996.
9. M. He, L. Yang, J. Zhang, and V. Vittal, "Spatio-temporal analysis for smart grids with wind generation integration," in *Computing, Networking and Communications (ICNC), 2013 International Conference on*, pp. 1107–1111, 2013.
10. D. Kugiumtzis and E. Bora-Senta, "Gaussian analysis of non-Gaussian time series," *Brussels Economic Review*, vol. 53, no. 2, pp. 295–322, 2010.
11. A. Kusiak, H. Zheng, and Z. Song, "Wind farm power prediction: a data-mining approach," *Wind Energy*, vol. 12, no. 3, pp. 275–293, 2009.
12. A. Lau and P. McSharry, "Approaches for multi-step density forecasts with application to aggregated wind power," *Ann. Appl. Stat.*, vol. 4, no. 3, pp. 1311–1341, 2010.
13. D. Lew, M. Milligan, G. Jordan, and R. Piwko, "The value of wind power forecasting," *NREL Conference Paper CP-5500-50814*, Apr. 2011.
14. P. Luh, Y. Yu, B. Zhang, E. Litvinov, T. Zheng, F. Zhao, J. Zhao, and C. Wang, "Grid Integration of Intermittent Wind Generation: a Markovian Approach." in press, *IEEE Trans. Smart Grids*.
15. H. Madsen, P. Pinson, G. Kariniotakis, H. A. Nielsen, and T. S. Nielsen, "Standardizing the performance evaluation of short-term wind power prediction models," *Wind Engineering*, vol. 29, no. 6, pp. 475–489, 2005.
16. C. Monteiro, H. Keko, R. Bessa, V. Miranda, A. Botterud, J. Wang, and G. Conzelmann, "A quick guide to wind power forecasting: state-of-the-art 2009." [Online] Available: http://www.dis.anl.gov/pubs/65614.pdf, 2009.
17. S. Murugesan, J. Zhang, and V. Vittal, "Finite state Markov chain model for wind generation forecast: a data-driven spatio-temporal approach," *Innovative Smart Grid Technologies, IEEE PES*, pp. 1–8, Jan. 2012.
18. NERC IVGTF Task 2.1 report: Variable Generation Power Forecasting for Operations. www.nerc.com/docs/pc/ivgtf/Task2-1(5.20).pdf, May 2010.
19. M. E. J. Newman, "Power laws, Pareto distributions and Zipf's law," *Contemporary Physics*, vol. 46, no. 5, pp. 323–351, 2005.
20. "NSF Initiative on Core Techniques and Technologies for Advancing Big Data Science & Engineering (BIGDATA)." Online [Available]: http://www.nsf.gov/pubs/2012/nsf12499/nsf12499.htm#toc.
21. G. Papaefthymiou and B. Klockl, "MCMC for wind power simulation," *IEEE Trans. on Energy Convers.*, vol. 23, pp. 234–240, Mar. 2008.

22. P. Pinson, "Very-short-term probabilistic forecasting of wind power with generalized logit-normal distributions," *Journal of the Royal Statistical Society: Series C (Applied Statistics)*, vol. 61, no. 4, pp. 555–576, 2012.

23. P. Pinson and G. Kariniotakis, "Conditional prediction intervals of wind power generation," *IEEE Trans. Power Syst.*, vol. 25, no. 4, pp. 1845–1856, 2010.

24. P. Pinson and H. Madsen, "Adaptive modelling and forecasting of offshore wind power fluctuations with Markov-switching autoregressive models," *Journal of Forecasting*, vol. 31, no. 4, pp. 281–313, 2012.

25. R. C. Prim, "Shortest connection networks and some generalizations," *Bell System Technical Journal*, vol. 36, pp. 1389–1401, 1957.

26. G. Samorodnitsky and M. Taqqu, *Stable Non-Gaussian Random Processes: Stochastic Models with Infinite Variance (Stochastic Modeling Series)*. Chapman and Hall/CRC, 1994.

27. S. Santoso, M. Negnevitsky, and N. Hatziargyriou, "Data mining and analysis techniques in wind power system applications: abridged," in *Power Engineering Society General Meeting, 2006. IEEE*, pp. 1–3, 2006.

28. E. S. Tackle and J. M. Brown, "Note on the use of Weibull statistics to characterize wind speed data," *Journal Appl. Meteorol*, vol. 17, pp. 556–559, 1978.

29. Q. Zhang and S. A. Kassam, "Finite-state Markov model for Rayleigh fading channels," *IEEE Trans. on Commun.*, vol. 47, pp. 1688–1692, Nov. 1999.

Chapter 3
Support Vector Machine Enhanced Markov Model for Short-Term Wind Power Forecast

Wind ramps introduce significant uncertainty in wind power generation. Reliable system operation, however, requires accurate detection and forecast of wind ramps, especially at high wind generation penetration levels. In this chapter, a support vector machine (SVM) enhanced Markov model for short-term wind power forecast is developed, taking into account not only wind ramps but also the diurnal non-stationarity and the seasonality of wind farm generation. Specifically, using the historical data of the wind turbine power outputs recorded at an actual wind farm, multiple finite-state Markov chains that take into account the diurnal non-stationarity and the seasonality of wind generation are first developed to model the "normal" fluctuations of wind generation. To deal with the wind ramp dynamics, an SVM is then employed, based on one key observation from the measurement data that wind ramps often occur with specific patterns. Next, the forecast by the SVM is integrated cohesively into the finite-state Markov chain. Based on the SVM enhanced Markov model, both (short-term) distributional forecasts and point forecasts are then derived.

3.1 Introduction

In order to meet the renewable portfolio standards (RPS) adopted by many states in the U.S., much effort is being invested to integrate renewable generation (particularly wind generation) into bulk power grids. Indeed, wind energy constitutes a significant portion of this renewable integration [1]. High penetration of wind generation, however, is expected to result in significant operational challenges [2], due to its non-dispatchability and variability. Reliable system operation (committing and dispatching conventional generation resources), however, requires accurate forecasts of future wind generation. Currently, forecasting wind generation for an individual wind farm typically has an error of 15–20 % [3], in sharp contrast to the load forecast. Wind generation forecast errors could result in either committing more conventional generation capacity than needed when the actual wind generation is above the forecast value, or using costly ancillary services and fast acting reserves when the actual wind generation is less than the forecast value. The latter situation becomes more

© The Author(s) 2014
L. Yang et al., *Spatio-Temporal Data Analytics for Wind Energy Integration*,
SpringerBriefs in Electrical and Computer Engineering, DOI 10.1007/978-3-319-12319-6_3

significant in the presence of wind ramps. Therefore, it is imperative to develop accurate forecast approaches for wind farm generation, especially for wind power ramps.

Different wind power forecast approaches have been developed to tackle the uncertainty of wind power in different time scales [4, 5]. For short-term wind power forecast, time-series models (e.g., auto-regressive models [6–8], Kalman filtering [9]), artificial intelligent methods (e.g., artificial neural networks [10], fuzzy neural networks [11], and support vector machines (SVMs) [12–14]) and data mining [15–17] have been proposed. Although these wind power forecast approaches have reasonably low average forecast errors over long-term wind measurement data from different wind farms, wind farm power forecast errors can still be high in the presence of wind power ramps (wind power increases or decreases largely within a limited time window). As shown in [18], large wind energy plants commonly experience changes in wind power output of 20 % of rated capacity over 1 h, and wind ramps occur approximately once every two days. Therefore, wind power ramp modeling is needed to develop accurate forecast approaches.

Wind power ramp forecasting has become an important topic of research [19]. Bossavy et al. [20] investigate the skill of numerical weather prediction ensembles to make probabilistic forecasts of ramp occurrence. Zheng et al. [17] propose a data-mining approach to predict wind farm power ramp rates, while Zareipour et al. [21] propose an SVM approach to predict the class of wind power ramp events. However, the diurnal non-stationarity and the seasonality of wind generation are not accounted for in these studies. Neglecting these essential features may result in high forecast errors. To address this issue, an SVM enhanced Markov model is developed in this chapter, in which the forecast model dynamically switches to a transition probability matrix of Markov chains based on the current observations, in order to capture the diurnal non-stationarity and the seasonality of wind generation. The idea of combining SVMs and Markov models has been proposed in [22–24] for pattern recognition. However, the proposed SVM enhanced Markov model is formulated in a very different manner from these works, considering the diurnal non-stationarity and the seasonality of wind generation.

State-of-the-art short-term wind power forecast approaches (see, e.g., in [4, 5]) are mostly for "point forecast" paradigms, i.e., forecasting the value of wind power at a future time. To efficiently integrate the wind generation, distributional forecasts are required to explicitly manage the uncertainty. One of the key advantages of distributional forecasts is that distributional forecasts enable system operators to maintain an acceptable level of risk. As shown in recent studies [25–27], stochastic scheduling of power systems based on distributional forecasts can improve the system efficiency, in terms of reducing system reserves. To this end, distributional forecasts are developed, e.g., [6–8, 28–31]. A detailed review of distributional forecast approaches can be found in [32]. However, the modeling of wind power ramps is not accounted for in these works. Based on the Markov-chain-based forecast model developed in Chap. 2, this chapter proposes an SVM enhanced Markov model to incorporate the modeling of wind power ramps, based on which both distributional and point forecasts are then derived. Compared with time-series models and data mining-based regression models, the proposed SVM enhanced Markov model does not suffer from high computational complexity, and can be computed offline. In particular, the transition probability matrix of Markov chains and the SVM classifiers are learned from historical data, without assuming any shape of the distribution of wind power. Moreover, the proposed SVM enhanced Markov model takes into account the diurnal non-stationarity and the seasonality of wind generation and the wind power ramp dynamics, whereas no existing studies consider all these issues in a unified framework.

In this chapter, an SVM enhanced Markov model for short-term wind power forecast is developed, taking into account not only wind ramps but also the diurnal non-stationarity and the seasonality of wind farm generation. Using both the statistical information of wind farm generation extracted from historical data and the recent past observations via the deployment of the SVM, the proposed SVM enhanced Markov model can significantly improve the forecast accuracy.

The main contents of this chapter are summarized below:

- The SVM enhanced Markov model for short-term wind power forecast is developed, which incorporates the design of finite-state Markov chains and wind power ramp modeling. Specifically, using the historical data of the wind turbine power outputs recorded at an actual wind farm, multiple finite-state Markov chains, taking into account the diurnal non-stationarity and the seasonality of wind generation, are first developed to capture the "normal" fluctuations of wind generation, and SVMs are then employed to capture the wind power ramp dynamics, based on one key observation from the measurement data that wind ramps often occur with specific patterns (reflected in the past observations).
- Built on the proposed SVM enhanced Markov model, short-term distributional and point forecasts of wind farm generation are developed. When forecasting the wind farm generation, the proposed short-term distributional and point forecasts dynamically switch to a transition probability matrix of Markov chains based on the current observations, and therefore can potentially capture the wind generation dynamics with ramp events. Based on wind measurement data from an actual wind farm, the proposed forecast approach demonstrates significantly improved forecast accuracy, compared with other approaches.

The rest of the chapter is organized as follows. In Sect. 3.2, the SVM enhanced Markov model is developed. In Sect. 3.3, numerical results are shown by using realistic wind measurement data from an actual wind farm. A summary of the chapter is provided in Sect. 3.4. The main notation used in the chapter is summarized in Table 3.1.

3.2 Support Vector Machine Enhanced Markov Model

Figure 3.1 illustrates the proposed SVM enhanced Markov model for short-term wind power forecast. The design of the SVM enhanced Markov model consists of two major steps: the design of finite-state Markov chains and the design of wind

Table 3.1 Summary of the main notation

Notation	Definition
d_{i_m}	Distance (the number of hops) between turbine i_m and root turbine r_m in class C_m
$\bar{m}$	Index of the reference MET
n_{ij}	Number of transitions from S_i to S_j encountered in the measurement data
q_k	Transition probability of from S_k to $S_{\hat{k}}$
$\mathbf{x}$	l-dimensional feature vector
y	Class label of wind power ramp rate
$\{\mathbf{x}, y\}$	Training sample
C_m	Wind turbine class m
M_k	Number of wind power ramp rates at state S_k
N_s	Number of states of Markov chain
N_t	Number of measurement data
$P_{i_m}(t)$	Power output of wind turbine i_m in class C_m
$P_{r_m}(t)$	Power output of root wind turbine r_m in class C_m
$P_{ag}(t)$	Aggregate power output of the wind farm
$P_{ag,m}(t)$	Aggregate power output of class C_m
$P_{ag,k}$	Representative generation level of state S_k
$P_{ag}^{\max}$	Rated capacity of the wind farm
$\hat{P}_{ag}$	Forecast of the wind farm generation
$\hat{P}_{ag}^{\mathrm{SVM}}$	Forecast of the wind farm generation by using the SVM model
$\Pr(A\|B)$	Probability of event A conditioned on event B
Q	Transition matrix of Markov chain
R_j^k	Wind power ramp rate of the jth class of wind power ramp rate at state S_k
S_k	State k in $\mathcal{S}$, $k \in \{1, \dots, N_s\}$
$S_{\hat{k}}$	Forecast state by using the SVM model
$W_{\bar{m}}(t)$	Wind speed measured at the reference MET $\bar{m}$
$\mathcal{S}$	State space of Markov chain
α_m	Linear regression coefficient for the parent-child turbine pairs of C_m
β_m	Linear regression coefficient for $W_m(t)$ as an affine function of $W_{\bar{m}}(t)$
γ	Regularization parameter
τ_k	Average duration of state S_k
σ^2	Parameter in the kernel function $K(\cdot)$
Γ	Wind farm generation level
$f_j(\cdot)$	Decision function of the jth class of wind power ramp rate
$F_{P_{ag}}(\cdot)$	Cumulative density function of farm aggregate wind generation

Table 3.1 (continued)

Notation	Definition
$G_{pw}(\cdot)$	"power curve" of the wind farm, which maps $W_{\tilde{m}}(t)$ to $P_{ag}(t)$
$K(\cdot)$	Kernel function of the SVM model
$L_{P_{ag}}(\cdot)$	Level crossing rate function
$U_m(\cdot)$	Power curve of class C_m, which maps $W_m(t)$ to $P_{i_m}(t)$, $\forall i_m \in C_m$
$\phi(\cdot)$	Mapping function

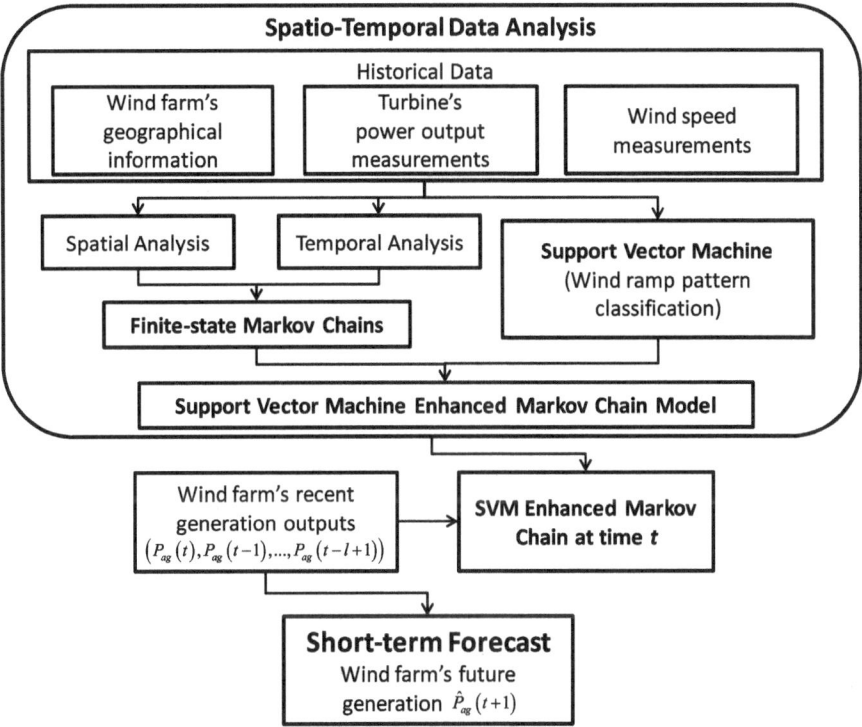

Fig. 3.1 Illustration of the support vector machine enhanced Markov model

ramp pattern classifiers based on the SVM. Specifically, finite-state Markov chains are developed by using a spatio-temporal analysis approach developed in Chap. 2: (1) the spatial analysis is carried out to study the spatial correlation of wind turbines' power output, aiming to obtain the cumulative density function of farm aggregate wind generation; (2) the temporal analysis is to capture the statistical distribution and temporal dynamics of aggregate wind farm generation. Then, wind ramp pattern classifiers based on the SVM are designed to deal with the wind ramp dynamics, and enhance the proposed finite-state Markov chains. The short-term wind power forecast is then derived based on the developed SVM enhanced Markov model. The details of the design of the SVM enhanced Markov model are described in the following.

3.2.1   Finite-State Markov Chains

The finite-state Markov chains are developed by using a spatio-temporal analysis approach proposed in Chap. 2. The key objective of the spatio-temporal analysis is to extract the statistical information of wind farm generation from historical data. A brief overview of the approach is first provided. The approach is then enhanced to account for wind power ramps.

3.2.1.1   Spatial Analysis

It is observed from the measurement data that the power outputs of wind turbines within the same wind farm can be quite different, even if the wind turbines are of the same class and physically located close to each other. Thus motivated, a graphical model to capture the spatial correlation between the power outputs from the wind turbines is developed, i.e., a minimum spanning tree is constructed based on graph theory. The spatial correlation between the individual wind turbines in each class C_m is determined by using a linear regression model, i.e., $P_{i_m}(t) = \alpha_m^{d_{i_m}} P_{r_m}(t)$, where $P_{i_m}(t)$ is the power output of wind turbine i_m in class C_m, $P_{r_m}(t)$ is the power output of root wind turbine r_m in the minimum spanning tree of class C_m, d_{i_m} is the distance (the number of hops) between turbine i_m and root turbine r_m, and α_m is a coefficient determined by the minimum mean square error (MMSE) principle as follows:

$$\alpha_m = \arg\min_{\alpha} \frac{1}{N_t} \sum_t \left(P_{ag,m}(t) - \sum_{i_m \in C_m} \alpha^{d_{i_m}} P_{r_m}(t) \right)^2, \tag{3.1}$$

where N_t is the number of measurement data, and $P_{ag,m}(t)$ is the aggregate power output of class C_m.

Based on (3.1) and the power curve $U_m(\cdot)$ of each class C_m, the aggregate power output of the wind farm $P_{ag}(t)$ can be characterized by using the wind speed $W_{\bar{m}}(t)$ measured at the reference meteorological tower (MET) $\bar{m}$ as follows:

$$P_{ag}(t) = \sum_m P_{ag,m}(t) = \sum_m \sum_{i_m \in C_m} \alpha_m^{d_{i_m}} U_m(\beta_m W_{\bar{m}}(t)) \triangleq G_{pw}(W_{\bar{m}}(t)), \tag{3.2}$$

where β_m is the correlation between the wind speed of the root wind turbine of class C_m and the reference MET $\bar{m}$, which can be obtained using the MMSE principle similarly as in (3.1) (see (2.3) in Chap. 2). $G_{pw}(\cdot)$ denotes the "power curve" of the wind farm, which maps $W_{\bar{m}}(t)$ to $P_{ag}(t)$. Therefore, the cumulative density function $F_{P_{ag}}$ of farm aggregate wind generation can be obtained from the cumulative density function $F_{W_{\bar{m}}}$ of $W_{\bar{m}}(t)$, i.e., $F_{P_{ag}}(\cdot) = F_{W_{\bar{m}}}(G_{pw}^{-1}(\cdot))$.

3.2.1.2   Temporal Analysis

The temporal correlation of wind generation is analyzed by using a finite state Markov chain model. A critical step is to capture the statistical distribution and temporal dynamics of aggregate wind farm generation $P_{ag}(t)$ using a Markov chain with the following characteristics:

- The Markov chain has N_s (N_s is finite) states. Let S denote the state space of the Markov chain. Specifically, state $S_k = [\Gamma_k, \Gamma_{k+1})$, $k \in \{1, \ldots, N_s\}$, corresponds to a specific range of generation levels with extreme values given by $\Gamma_1 = 0$ and $\Gamma_{N_s+1} = P_{ag}^{max}$;
- The Markov chain is discrete-time and of order 1.

The above characteristics are adopted to make the Markov chains practical for forecasting applications, so that forecast is made based on the most recent data only.

The objective of the Markov chain design is to determine the generation levels Γ_k ($k = 1, \ldots, N_s + 1$) that define the states, the transition matrix Q, and the representative generation level $P_{ag,k}$ for each state k. The procedure developed in [33] is utilized to design the state space. Define the quantity τ_k as the average duration that P_{ag} stays in state S_k,

$$\tau_k = \frac{F_{P_{ag}}(\Gamma_{k+1}) - F_{P_{ag}}(\Gamma_k)}{L_{P_{ag}}(\Gamma_{k+1}) + L_{P_{ag}}(\Gamma_k)}, \tag{3.3}$$

where $L_{P_{ag}}(\Gamma)$ denotes the level crossing rate (the number of times per unit time that the farm aggregate power process $\mathcal{P}_w$ crosses Γ) for the farm aggregate power Γ ($\Gamma \geq 0$) [33].

By using the method in [33], a constant τ is introduced to find the $N_s - 1$ variables $\{\Gamma_2, \Gamma_3, \ldots, \Gamma_{N_s}\}$, i.e., solving (3.3) with $\tau_k = \tau$, $\forall k \in \{1, \ldots, N_s - 1\}$. Then, the transition probability matrix could be easily obtained by using the data, as follows:

$$Q_{i,j} = \frac{n_{ij}}{\sum_{k=1}^{N_s} n_{ik}}, \tag{3.4}$$

where n_{ij} is the number of transitions from S_i to S_j. Accordingly, the representative generation level for each state S_k can be obtained by

$$P_{ag,k} = \frac{\int_{\Gamma_k}^{\Gamma_{k+1}} x f_{P_{ag}}(x) dx}{F_{P_{ag}}(\Gamma_{k+1}) - F_{P_{ag}}(\Gamma_k)}, \tag{3.5}$$

where $f_{P_{ag}}(\cdot)$ is the probability density function of P_{ag}.

Note that due to the diurnal non-stationarity and the seasonality of wind generation, the Markov chain is non-stationary. The seasonality is tackled by designing the forecast model for each month individually. The diurnal non-stationarity is handled by identifying an epoch such that the wind generation exhibits stationary behavior within every such epoch and designing a forecast model for each of these epochs separately. The detailed design and discussions of the Markov chains can be found in Chap. 2.

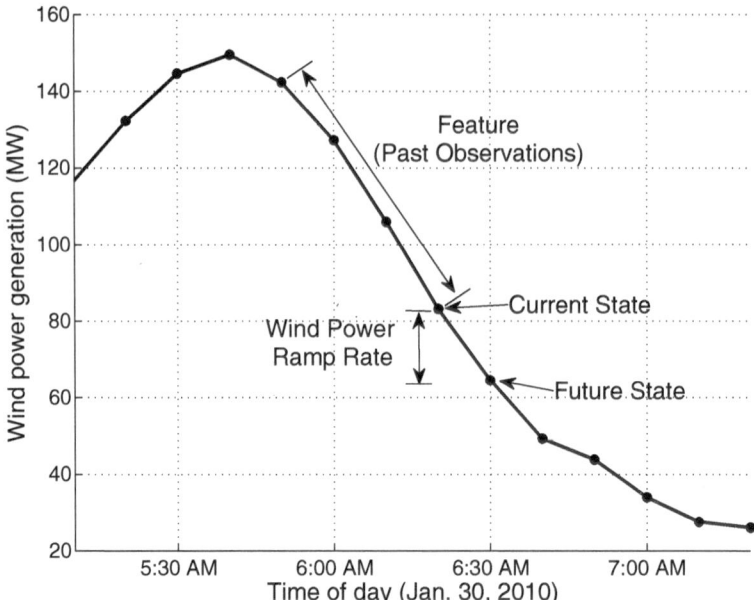

Fig. 3.2 Strong correlation between the past observation and the wind power ramp rate

3.2.2 Wind Power Ramp Classification using SVM

The proposed finite-state Markov chains in the previous section can capture the small fluctuations of wind generation, as the Markov chains focus on the transitions among neighboring states.[1] As shown in [18], large wind energy plants commonly experience changes in wind power output of 20 % of rated capacity over 1 h, and wind ramps occur approximately once every two days. Short-term wind power forecasts based on only the proposed finite-state Markov chains may still have significant uncertainty and error, due to wind power ramps. To capture the wind ramp dynamics, an SVM is employed to enhance the proposed finite-state Markov chains, based on one key observation from the measurement data that wind ramps often occur with specific patterns (past observations) that are highly correlated with the wind power ramp rates. For example, in Fig. 3.2 when a wind ramp-down event occurs, the wind power ramp rate can be forecasted from the recent past observations (patterns). As SVMs can effectively capture the relationship between such patterns and the wind power ramp rates, we propose to employ SVMs to capture the wind ramp dynamics, in order to enhance the proposed finite-state Markov chains.

---

[1] Examples of the designed state space and the corresponding transition matrix can be found in Chap. 2.

In SVMs, the past observations are called *features*, and the wind power ramp rate (the wind power difference between the current state and the future state as shown in Fig. 3.2) is called a *class*. As the wind power is quantized into multiple states based on the state space design of the proposed finite-state Markov chains, the classes, i.e., wind power ramp rates, can be determined using the representative generation level for each state. Note that the classes are different for different epochs in different months, as different Makrov chains are designed to account for the diurnal non-stationarity and the seasonality of wind generation. The number of classes is the same as the number of possible future states, which can be determined from the measurement data.

In SVMs, the basic idea is to map a feature into a higher-dimensional feature space via a nonlinear mapping, based on which the SVM classifier predicts its class, i.e., the wind power ramp rate. The objective is to obtain a characterization of the classifier based on the historical data. Note that the characteristics of wind power ramps can be different for different states in different Markov chains, due to the diurnal non-stationarity and the seasonality of wind generation. Therefore, for each state in each Markov chain, a unique SVM classifier is developed.

Since the number of wind power ramp classes (i.e., the possible values of the wind power ramp rate) is greater than two, a multi-class classifier is needed. In this study, the one-against-all method [34] is used to construct multiple binary SVM classifiers with the jth one separating class j from all the remaining classes. In what follows, let M_k be the number of wind power ramp classes at state S_k and R_j^k be the wind power ramp rate of the jth class. Here, the index for each Markov chain is omitted for notational simplicity. Specifically, for each state S_k, let $\{\mathbf{x}_1, y_1\}, \dots, \{\mathbf{x}_{N_k}, y_{N_k}\}$ be N_k training samples, where $\mathbf{x}_i \in \mathbb{R}^l$ is an l-dimensional feature vector (l observations) and $y_i \in \{1, \dots, M_k\}$ is the corresponding class label. In the one-against-all method, the jth SVM is trained with all the training samples of the jth class with positive labels, and the others with negative labels, which requires the solution to the following quadratic optimization problem that yields the jth decision function $f_j(\mathbf{x}) = w_j^T \phi(\mathbf{x}) + b_j$, where $w_j \in \mathbb{R}^l$, b_j is a scalar, and $\phi(\mathbf{x})$ maps $\mathbf{x}$ into a higher-dimensional space.

$$
\begin{aligned}
\text{minimize} \quad & \tfrac{1}{2} w_j^T w_j + \gamma_j \sum_{i=1}^{N_k} \xi_i^j \\
\text{subject to} \quad & \tilde{y}_i (w_j^T \phi(\mathbf{x}_i) + b_j) \geq 1 - \xi_i^j, \\
& \xi_i^j \geq 0, \quad i = 1, \dots, N_k, \\
\text{variables} \quad & \{w_j, b, \xi^j\},
\end{aligned}
\tag{3.6}
$$

where $\tilde{y}_i = 1$ if $y_i = j$ and $\tilde{y}_i = -1$ otherwise, and $\gamma_j > 0$ is the regularization parameter. After solving (3.6) for each class, there are M_k decision functions associated with each class.

At the classification phase, a sample $\mathbf{x}$ is classified to belong to class j^*, if $f_{j^*}(\mathbf{x})$ produces the largest value:

$$
j^* = \operatorname*{arg\,max}_{j \in \{1, \dots, M_k\}} f_j(\mathbf{x}) = \operatorname*{arg\,max}_{j \in \{1, \dots, M_k\}} (w_j^T \phi(\mathbf{x}) + b_j).
\tag{3.7}
$$

Accordingly, the forecast of the wind farm generation by using the SVM is

$$\hat{P}_{ag}^{\text{SVM}}(t+1) = P_{ag}(t) + R_{j*}^k, \tag{3.8}$$

where $P_{ag}(t) \in [\Gamma_k, \Gamma_{k+1})$ is the current observed wind farm generation. The corresponding forecast state $S_{\hat{k}}$ is the state satisfying $\hat{P}_{ag}^{\text{SVM}}(t+1) \in [\Gamma_{\hat{k}}, \Gamma_{\hat{k}+1})$.

The forecast state $S_{\hat{k}}$ using SVM can be used to enhance the transition matrix of the proposed finite-state Markov chains, and therefore the wind ramp dynamics can be captured.

3.2.3 SVM Enhanced Markov Model for Wind Power Forecast

In this section, the SVM enhanced Markov model is developed by integrating the SVM model into the Markov chain model. The basic idea is to integrate the forecast state using SVM into the transition matrix of the Markov chain. This is done by using the estimated forecast accuracy of the SVM model. Specifically, the estimated forecast accuracy of the SVM model is used as the transition probability from the current state to the forecast state $S_{\hat{k}}$ given by (3.8), which can be obtained by using the cross-validation approach [35]. In the cross-validation approach, the training set is divided into v subsets of equal size. Sequentially each subset is tested using the classifier trained on the remaining $v - 1$ subsets. Thus, each instance of the whole training set is predicted once, so the cross-validation accuracy is the percentage of data which are correctly classified. Let q_k denote the transition probability from the current state S_k to the forecast state $S_{\hat{k}}$ given by (3.8). The distributional forecast is then given by

$$\Pr(P_{ag}(t+1) = P_{ag,j} | S(t) = S_k, \mathbf{x}(t)) = \begin{cases} q_k, & \text{if } j = \hat{k}; \\ (1 - q_k)\dfrac{Q_{k,j}}{\sum_{l \neq \hat{k}} Q_{k,l}}, & \text{otherwise,} \end{cases} \tag{3.9}$$

where $S(t)$ is the state of the Markov chain at time t and $\mathbf{x}(t) \in \mathbb{R}^l$ is the feature vector consisting of the latest l observations till time t. Note that the forecast state $S_{\hat{k}}$ depends on the observations $\mathbf{x}(t)$ at time t, which makes the transition distribution (3.9) capable of capturing the dynamics of the wind power ramps, as the proposed SVM enhanced Markov model dynamically switches to a transition probability matrix of Markov chains based on the current observations.

From the distributional forecast (3.9), a point forecast can be obtained as

$$\hat{P}_{ag}(t+1) = P_{ag,\hat{k}}q_k + \sum_{j \neq \hat{k}} P_{ag,j}(1 - q_k)\frac{Q_{k,j}}{\sum_{l \neq \hat{k}} Q_{k,l}}, \tag{3.10}$$

which is the conditional mean of the SVM enhanced Markov chain conditioned on the current state $S(t)$ and the observations $\mathbf{x}(t)$.

Remarks:

- The forecasts (3.9) and (3.10) can capture the wind power dynamics by using a hybrid model based on Markov chains and SVMs. When wind ramp events occur, q_k would capture the transition from the current state to some faraway state $\hat{k}$ and closely track the dynamics of wind farm power.
- The proposed model is trained offline and can be implemented in parallel, when computing the state space and the transition matrix of each Markov chain, since each Markov chain is obtained based on data in different epochs and months, in order to take into account the diurnal non-stationarity and the seasonality of wind farm generation. After obtaining the Markov chains, the SVM classifiers can also be computed in parallel, because they are trained by using different training data. Therefore, the Markov chains and the SVM classifiers can be obtained efficiently.

3.3   Case Studies

3.3.1   Data Description

The data used in the case studies is from a large wind farm with a rated capacity of $P_{ag}^{\max} = 300.5$ MW, which is the same as in Chap. 2. There are two classes of wind turbines in this wind farm, one class with 53 turbines and the other class with 221 turbines. For each class C_m, a meteorological tower (MET) is deployed and co-located with a wind turbine, i.e., the root turbine r_m. The instantaneous power outputs of all wind turbines and the wind speeds measured at all METs are recorded every 10 min for the years 2009 and 2010. In the case studies, the data of year 2009 is used to train the Markov chains and the SVM classifiers, and the data of year 2010 is used to test the forecast accuracy of the proposed method. The SVM classifiers are trained by using the LIBSVM toolbox [36].

3.3.2   SVM Parameter Selection

3.3.2.1   Kernel Function

In the SVM, the mapping function $\phi(\mathbf{x})$ can be implicitly defined by introducing the so-called kernel function $K(\mathbf{x}_{i_1}, \mathbf{x}_{i_2})$, which computes the inner product of vectors $\phi(\mathbf{x}_{i_1})$ and $\phi(\mathbf{x}_{i_2})$. In this paper, the radial basis function [37] is used,

$$K(\mathbf{x}_{i_1}, \mathbf{x}_{i_2}) = \exp\left(-\frac{\|\mathbf{x}_{i_1} - \mathbf{x}_{i_2}\|^2}{\sigma_j^2}\right), \tag{3.11}$$

where σ_j^2 is a predetermined parameter and may be different for different classifiers.

3.3.2.2   Grid Search

The parameters γ_j and σ_j^2 of the SVM are determined based on the grid-search approach [35]. The basic idea of the grid-search approach is to select γ_j and σ_j^2 from a grid that is formed by exponentially growing sequences of γ_j and σ_j^2 (for example, $\gamma_j = 2^{-5}, 2^{-3}, \dots, 2^{15}$ and $\sigma_j^2 = 2^{-15}, 2^{-13}, \dots, 2^3$). Various pairs of γ_j and σ_j^2 from a grid are tested and the one with the best cross-validation accuracy is picked.

3.3.3   Short-Term Wind Power Forecast

The proposed spatio-temporal analysis approach is applied to the measurement data of year 2009. To tackle the seasonality and the diurnal non-stationarity, a unique Markov chain for each epoch of a day (3 h for the wind farm considered here) in each month is developed. For example, in the 6AM-9AM epochs of January, a unique Markov chain is developed and will be used to forecast the wind farm generation. As a result, in each month, 8 Markov chains are obtained, each for a 3-h epoch. For each state in each Markov chain, a multi-class classifier is constructed by using the one-against-all method described in Sect. 3.2.2 with the SVM parameters determined in Sect. 3.3.2. Using a PC with a 2.4 GHz Intel Core i3 processor and 4 GB RAM, each Markov chain can be computed in about 20 s and each SVM classifier can be computed in about 10 s. We would like to emphasize that the training is done offline, and therefore the online computational complexity is low.

Then, the corresponding measurement data of year 2010 is used to test the performance of the proposed SVM enhanced Markov forecast approach.

3.3.3.1   Impact of Feature Dimension

Figure 3.3 compares the point forecast errors given by different values of the feature dimension l (the number of past observations) by using the data of year 2010, where the forecast error is quantified by mean absolute error (MAE),

$$\text{MAE} = \frac{1}{N_t} \sum_t |P_{ag}(t) - \hat{P}_{ag}(t)|. \tag{3.12}$$

As shown in Fig. 3.3, the MAE decreases dramatically with the increase of the feature dimension up to $l = 5$ and increases thereafter. This is mainly due to the diurnal non-stationarity and the seasonality of wind power. Intuitively speaking, if the feature dimension, i.e., the number of previous data samples used as the input of the SVM, is too high, bad features would reduce the classification accuracy, whereas if the feature dimension is low, there would not be enough features to capture the characteristics of wind power dynamics. Therefore, improper selection of the feature dimension would lead to poor classification in the SVM. The selection of the feature

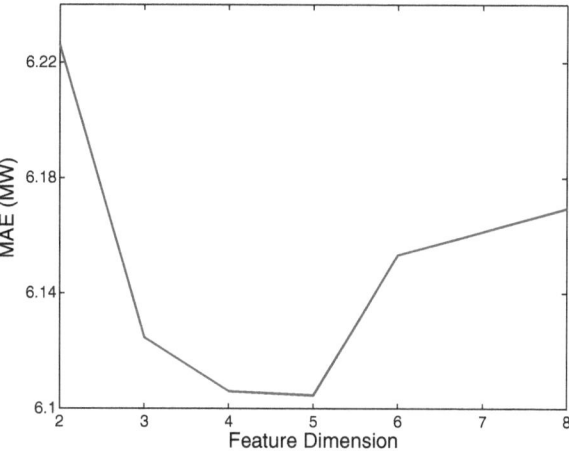

Fig. 3.3 MAE as the function of the feature dimension

dimension can be based on the autocorrelation coefficients of the data samples as suggested in [13]. Based on Fig. 3.3, $l = 5$ is selected as the feature dimension and is used in the following simulations.

3.3.3.2 Distributional Forecasts

The continuous rank probability score (CRPS) is utilized to evaluate the performance of the proposed distributional forecasts [7]. The CRPS is defined as:

$$\text{CRPS} = \frac{1}{N_t} \sum_t \int_0^{P_{ag}^{\max}} \left(\hat{F}_t(x) - H(x - P_{ag}(t)) \right)^2 dx,$$

where $\hat{F}_t(x)$ is the cumulative density function (CDF) obtained by using the distributional forecast (3.9), and $H(x - P_{ag}(t))$ is the unit step function that equals to 0 if $x < P_{ag}(t)$ and 1 otherwise. In principle, the higher the CRPS, the less accurate the distributional forecast is.

Table 3.2 provides the CRPS of the proposed distributional forecast using the actual data of year 2010. For comparison, the CRPSs given by the distributional forecasts based on the Markov-chain-based forecast model in Chap. 2 and AR models are also provided, where two AR models with a truncated Gaussian distribution [7] and a truncated log-normal distribution [8] are considered. The AR model with a Gaussian distribution is adopted from [7] by considering one regime, and then the support of the Gaussian distribution is truncated into $[0, P_{ag}^{\max}]$. The procedure for building AR models with truncated log-normal distributions can be found in [8]. Specifically, the order of the AR models are determined by using the partial autocorrelation functions of the wind power time series [38]. Then, the recursive least square algorithm [38] is applied to calculate the regressive coefficients, the predicted wind power $\hat{P}_{ag}(t)$ (the point forecast of the AR model), and the variance

Table 3.2 CRPS of distributional forecasts (as a percentage of the nominal capacity $P_{ag}^{\max}$) over the data of year 2010 . "SVM-MC" denotes the proposed SVM enhanced Markov model. "MC" denotes the Markov-chain-based forecast model

	SVM-MC	MC	AR(Gaussian)	AR(Log-normal)
CRPS (%)	1.91	2.03	2.29	2.18

Table 3.3 CRPS of distributional forecasts (as a percentage of the nominal capacity $P_{ag}^{\max}$) over all ramp events in year 2010 with $P_{th} = 15$ MW. "SVM-MC" denotes the proposed SVM enhanced Markov model. "MC" denotes the Markov-chain-based forecast model

	SVM-MC	MC	AR(Gaussian)	AR(Log-normal)
CRPS (%)	3.48	4.03	3.46	3.36

of innovation $\mathbf{C}$. Finally, by using $\hat{P}_{ag}(t)$ as the mean and $\mathbf{C}$ as the variance of a Gaussian distribution or a log-normal distribution which is truncated into $[0, P_{ag}^{\max}]$, the wind power distributional forecasts can be obtained.

As shown in Table 3.2, the improvements of the proposed distributional forecast in terms of the CRPS criterion are significant: 5.91 % lower than the Markov-chain-based forecast, 16.59 % lower than the AR-Gaussian-based forecast, and 12.39 % lower than the AR-Log-normal-based forecast, respectively. The reason for the improvements of the proposed distributional forecast is two-fold: (1) The proposed SVM enhanced Markov model does not assume the shape of the distribution (and thus can be regarded as "non-parametric" distributional forecasts [4]); (2) The deployment of the SVMs can effectively capture the wind ramp dynamics, and therefore reduce the forecast errors.

Table 3.3 provides the CRPS of the proposed distributional forecast for all ramp events in year 2010. The definition of ramp events in [19] is used to identify ramp events. In the simulation, a ramp event is considered to occur if $|P_{ag}(t+1) - P_{ag}(t)| > P_{th}$, where P_{th} is a predefined threshold. In Table 3.3, P_{th} is set to be 15 MW, which is equivalent to 5 % of the nominal capacity. As shown in Table 3.3, by employing the SVM to capture the wind power ramps, the proposed distributional forecast improves the Markov-chain-based forecast by 13.6 %, which demonstrates the effectiveness of using the SVM to capture the wind power ramps. It is also observed that the proposed distributional forecast and the AR-based forecasts are comparable. Note that the proposed models are trained in an offline manner with the data of year 2009, whereas the AR-based models are trained in an online manner. Therefore, the online computational complexity of the proposed distributional forecast is much less than the AR-based forecasts.

To further examine the performance of the proposed SVM enhanced Markov model, the median and percentiles of the CRPS values over the data points for each month and each epoch are illustrated in Figs. 3.4 and 3.5, respectively. In Figs. 3.4 and 3.5, the middle bar in a box represents the median value of the CRPS values over all data points that fall into a specific month or epoch. The top edge and bottom edge of a box represent the 25th and 75th percentiles, respectively. The top bar and bottom bar correspond to the extremes calculated from 1.5 interquartile ranges. It is

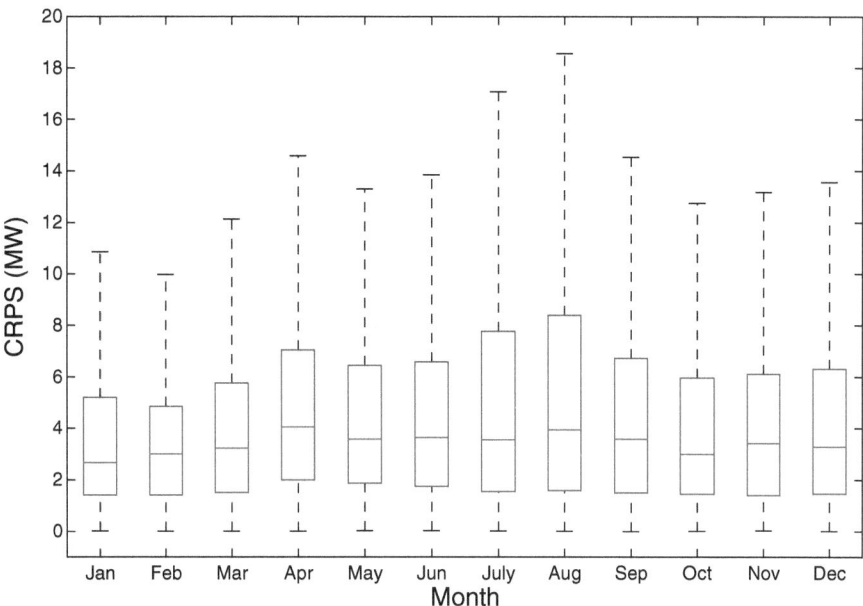

Fig. 3.4 Statistics of CRPS for 12 months of year 2010

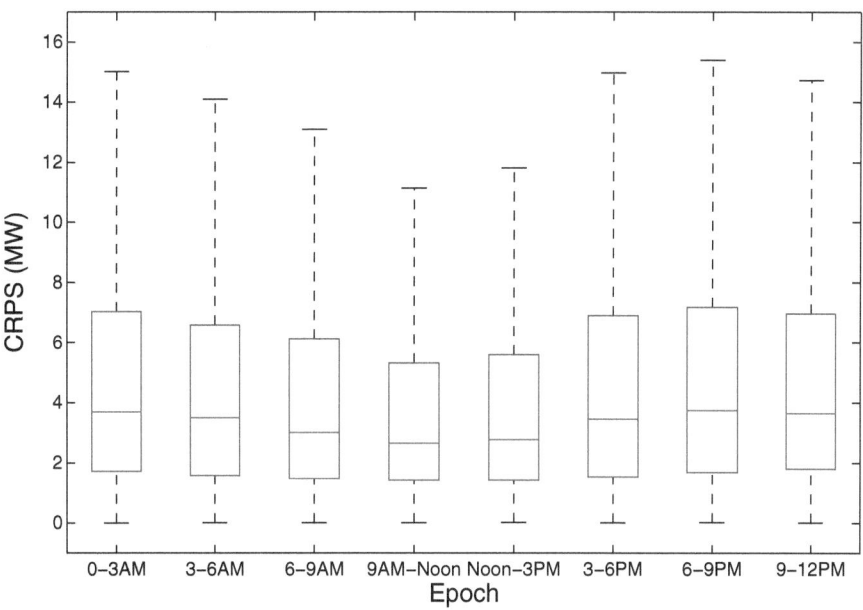

Fig. 3.5 Statistics of CRPS for 8 epochs of year 2010

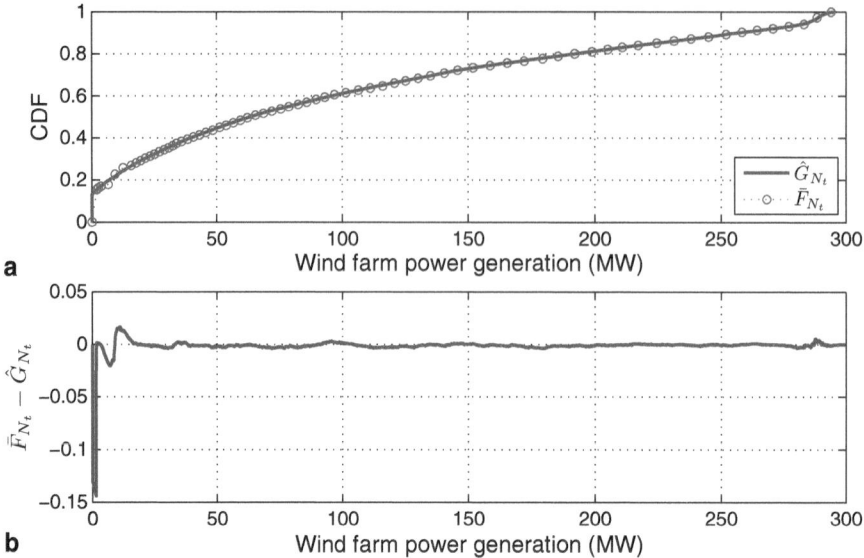

Fig. 3.6 Marginal calibration for the SVM enhanced Markov model, using the data of year 2010. **a** The *upper figure* compares the average predictive CDF with the empirical CDF of the actual wind generation; **b** The *lower figure* illustrates the difference of two CDFs

observed from Figs. 3.4 and 3.5 that the deviations of the CRPS values are slightly higher across the summer season and during afternoon and night epochs, when the wind power ramps frequently occur.

Next, a marginal calibration test [39] is carried out to examine the reliability of the SVM enhanced Markov model. Marginal calibration examines the equality of the wind generation forecast and the actual wind generation. To assess marginal calibration, we follow the procedure developed in [39] by comparing the average predictive CDF,

$$\overline{F}_{N_t}(x) = \frac{1}{N_t} \sum_t \hat{F}_t(x), \tag{3.13}$$

with the empirical CDF of the actual wind generation,

$$\hat{G}_{N_t}(x) = \frac{1}{N_t} \sum_t \mathbf{1}(x_t \leq x), \tag{3.14}$$

where $\mathbf{1}(x_t \leq x)$ denotes the indicator function that equals to 1 if $x_t \leq x$ and 0 otherwise. Figure 3.6 illustrates the marginal calibration test of the distributional forecast using the SVM enhanced Markov model. It is observed that the average predictive CDF using the SVM enhanced Markov model matches the empirical CDF of the actual wind generation and the difference of the two CDFs is relatively large when the wind generation is at a low level. Since wind generation is quite often at a medium level, the training data may not capture the dynamics of wind generation at a low level, which results in this relatively large difference at a low level.

Further, to better illustrate the proposed distributional forecast, Fig. 3.7 provides two episodes of prediction intervals. January 23rd, 2010, is chosen for illustration, in which the wind power fluctuates significantly with the highest average ramp rate of 15.44 MW per 10 min. Another extreme day, January 30th, 2010 is chosen for illustration, in which the wind power generation changes dramatically during the period 3:00AM-9:00AM by more than 100 MW. As illustrated in Fig. 3.7, the actual wind farm generation lies in the prediction interval obtained from the proposed distributional forecast, despite the sharp ramps.

3.3.3.3 Point Forecasts

We compare the forecast errors of the proposed approach with three point forecast approaches,

- Persistence forecast: $\hat{P}_{ag}(t+1) = P_{ag}(t)$.
- Forecast with the adaptive AR model.
- The Markov-chain-based forecast developed in Chap. 2.

Forecast errors are quantified by MAE (3.12) and root mean square error (RMSE),

$$\text{RMSE} = \sqrt{\frac{1}{N_t} \sum_t |P_{ag}(t) - \hat{P}_{ag}(t)|^2}. \tag{3.15}$$

The test results over the data of the entire year 2010 and all ramp events of the entire year 2010 with $P_{th} = 15$ MW are provided in Tables 3.4 and 3.5, respectively. It is observed that the proposed SVM enhanced Markov model outperforms the other approaches. From Table 3.4, the improvements of the proposed approach are significant: (1) in terms of MAE, 5.45 % lower than the Markov-chain-based forecast, 10.34 % lower than the persistence forecast, and 1.89 % lower than the AR-based forecast; (2) in terms of RMSE, 4.27 % lower than the Markov-chain-based forecast, 10.71 % lower than the persistence forecast, and 1.47 % lower than the AR-based forecast. From Table 3.5, it is observed that there is a clear advantage of using the proposed SVM enhanced Markov model, as the improvement over the persistence forecast is up to 30 % in terms of MAE and 21 % in terms of RMSE during wind ramp events. It is also observed that using the SVM, the proposed approach can improve the Markov-chain-based forecast.

Furthermore, the statistics of the absolute error of the point forecasts by the proposed SVM enhanced Markov model over different months and epochs are shown in Figs. 3.8 and 3.9, respectively. It can be seen from Figs. 3.8 and 3.9 that the statistics of absolute error are similar for 12 months and 8 epochs, indicating that the proposed SVM enhanced Markov model performs consistently across the entire year.

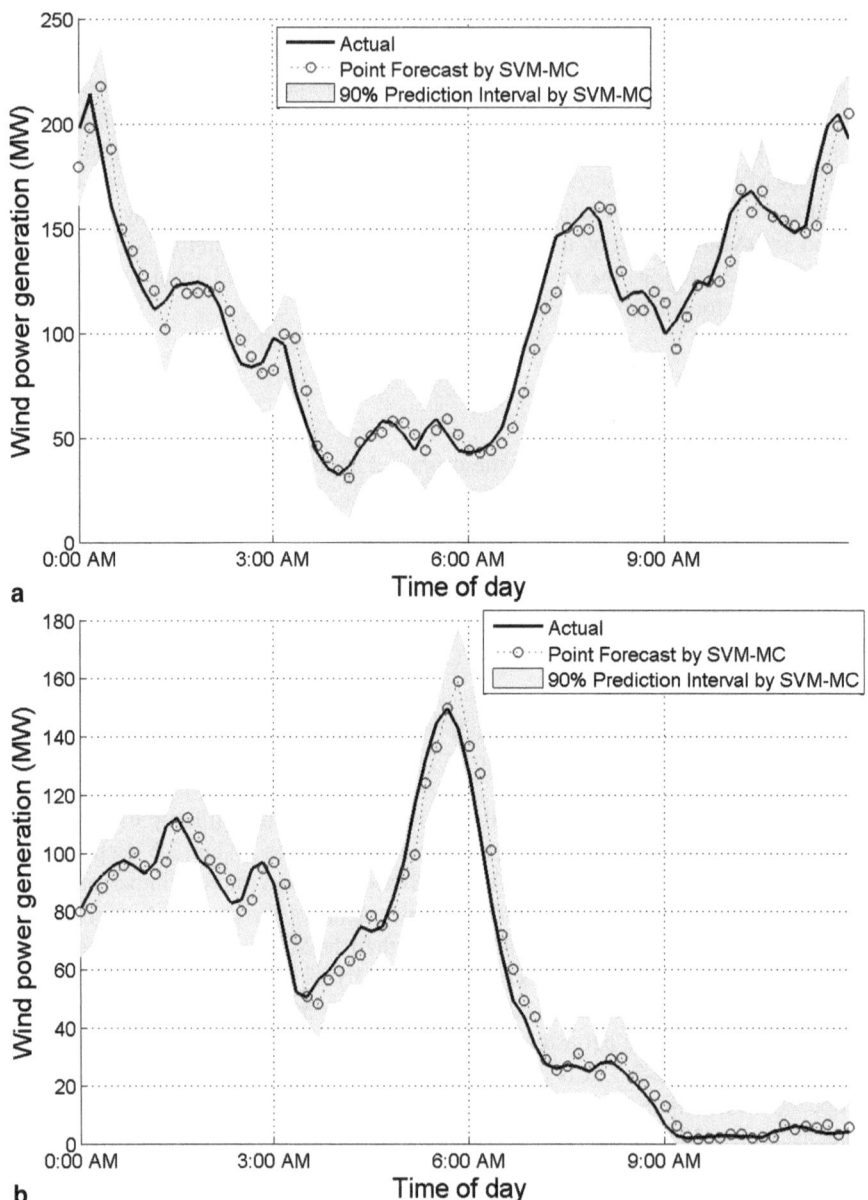

Fig. 3.7 Ten minutes distributional forecasts

Table 3.4 Comparison of different forecast approaches over the data of the entire year 2010, where "SVM-MC" denotes the proposed SVM enhanced Markov model, and "MC" denotes the Markov-chain-based forecast model. MAE and RMSE are normalized by the nominal capacity P_{ag}^{max}

Error	SVM-MC	MC	Persistence	AR
MAE (%)	2.08	2.20	2.32	2.12
RMSE (%)	3.36	3.51	3.72	3.41

Table 3.5 Comparison of different forecast approaches over all ramp events of the entire year 2010 with $P_{th} = 15$ MW. "SVM-MC" denotes the proposed SVM enhanced Markov model, and "MC" denotes the Markov-chain-based forecast model. MAE and RMSE are normalized by the nominal capacity P_{ag}^{max}

Error	SVM-MC	MC	Persistence	AR
MAE(%)	4.01	4.04	5.72	4.25
RMSE (%)	4.56	4.59	5.74	5.31

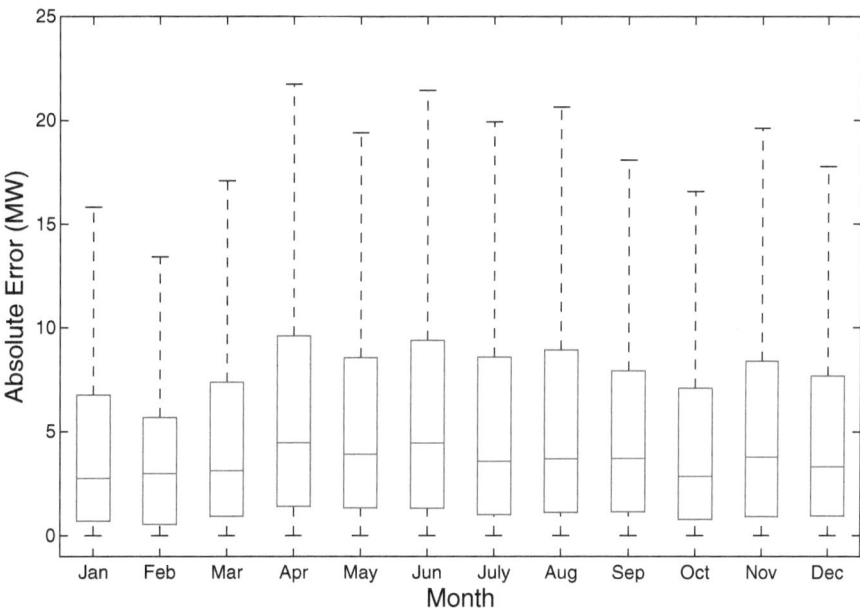

Fig. 3.8 Statistics of absolute error for 12 months of year 2010

From the results presented above, it can be seen that the proposed SVM enhanced Markov model can provide both distributional and point forecasts more accurate than those given by the other approaches. The proposed SVM enhanced Markov forecast is based on not only the statistical information of wind farm generation extracted from historical data, but also recent past observations. In contrast, the Markov-chain-based forecast and the AR-based approaches use only part of these information. Specifically,

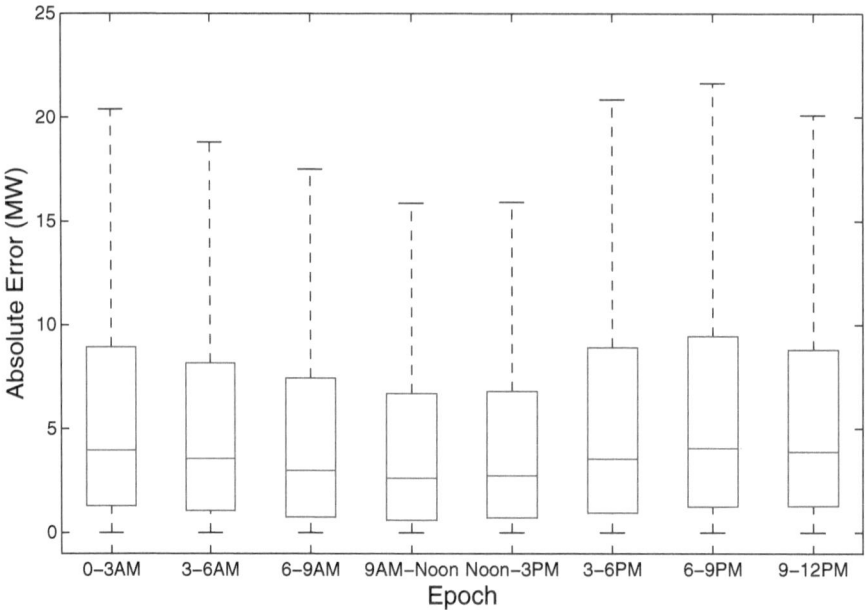

Fig. 3.9 Statistics of absolute error for 8 epochs of year 2010

the Markov-chain-based forecast does not consider the wind ramp patterns, while the AR-based approaches use only recent past observations. Therefore, the improvement of the proposed SVM enhanced Markov approach over other approaches can be attributed to: (1) the rigorous design of Markov chains and transition probabilities, which extract from historical data the statistical information of wind farm generation; (2) the deployment of the SVM, which can effectively capture the wind ramp dynamics based on recent past observations. It is also worth emphasizing that the proposed SVM enhanced Markov models are trained offline and therefore the online computational complexity is much less than the adaptive AR-based approaches that require online training to calculate the regressive coefficients.

3.4 Summary

In this chapter, the support vector machine (SVM) enhanced Markov model for short-term wind power forecast is developed. This approach takes into account not only wind ramps but also the diurnal non-stationarity and the seasonality of wind farm generation. Specifically, using the historical data of the wind turbine power outputs, multiple finite-state Markov chains that take into account the diurnal non-stationarity and the seasonality of wind generation are first developed to capture the small fluctuations of wind generation. To capture the wind ramp dynamics, the

SVM is employed, based on one key observation from the measurement data that wind ramps often occur with specific patterns. Besides, to capture the diurnal non-stationarity and the seasonality of wind generation, multiple SVM classifiers are used and each of them is associated with each state in each Markov chain. Then, the forecast state by the SVM is integrated into each finite-state Markov chain. Based on the SVM enhanced Markov model, short-term distributional forecasts and point forecasts are then derived. Numerical test results demonstrate the significant improved accuracy of the proposed forecast approach.

In future work, it is of great interest to generalize the proposed Markov-chain based forecast model to multi-step ahead wind power forecasting and study stochastic unit commitment and dispatch problems in a Markov-chain based stochastic optimization framework.

References

1. K. Cory and B. Swezey, "Renewable portfolio standards in the states: balancing goals and implementation strategies." NREL Technical Report TP-670-41409, Dec. 2007.
2. L. Xie, P. Carvalho, L. Ferreira, J. Liu, B. Krogh, N. Popli, and M. Ilic, "Wind integration in power systems: operational challenges and possible solutions," *Proc. IEEE*, vol. 99, no. 1, pp. 214–232, 2011.
3. D. Lew, M. Milligan, G. Jordan, and R. Piwko, "The value of wind power forecasting," *NREL Conference Paper CP-5500-50814*, Apr. 2011.
4. G. Giebel, R. Brownsword, G. Kariniotakis, M. Denhard, and C. Draxl, *The State of the Art in Short-Term Prediction of Wind Power—A Literature Overview, 2nd Edition*. ANEMOS.plus, 2011. [Online] Available: http://www.anemos-plus.eu/images/pubs/deliverables/aplus.deliverable_d1.2.stp_sota_v1.1.pdf.
5. C. Monteiro, H. Keko, R. Bessa, V. Miranda, A. Botterud, J. Wang, G. Conzelmann, and I. Porto, *A quick guide to wind power forecasting : state-of-the-art 2009*. 2009. [Online] Available: http://www.dis.anl.gov/pubs/65614.pdf.
6. P. Pinson and H. Madsen, "Probabilistic Forecasting of Wind Power at the Minute Time-Scale with Markov-Switching Autoregressive Models," in *Probabilistic Methods Applied to Power Systems, 2008. PMAPS '08. Proceedings of the 10th International Conference on*, pp. 1–8, May 2008.
7. P. Pinson and H. Madsen, "Adaptive modelling and forecasting of offshore wind power fluctuations with Markov-switching autoregressive models," *Journal of Forecasting*, vol. 31, no. 4, pp. 281–313, 2012.
8. P. Pinson, "Very-short-term probabilistic forecasting of wind power with generalized logit-normal distributions," *Journal of the Royal Statistical Society: Series C (Applied Statistics)*, vol. 61, no. 4, pp. 555–576, 2012.
9. T. S. Nielsen, H. Madsen, H. A. Nielsen, P. Pinson, G. Kariniotakis, N. Siebert, I. Marti, M. Lange, U. Focken, L. von Bremen, P. Louka, G. Kallos, and G. Galanis, "Short-term Wind Power Forecasting Using Advanced Statistical Methods," in *Proceedings of European wind energy conference*, (Athens, Greece), pp. 1–9, 2006.
10. J. Catalao, H. M. I. Pousinho, and V. Mendes, "Hybrid wavelet-pso-anfis approach for short-term wind power forecasting in portugal," *IEEE Transactions on Sustainable Energy*, vol. 2, no. 1, pp. 50–59, 2011.
11. P. Pinson and G. Kariniotakis, "Wind power forecasting using fuzzy neural networks enhanced with on-line prediction risk assessment," in *Power Tech Conference Proceedings, 2003 IEEE Bologna*, vol. 2, 2003.

12. M. Mohandes, T. Halawani, S. Rehman, and A. A. Hussain, "Support vector machines for wind speed prediction," *Renewable Energy*, vol. 29, no. 6, pp. 939–947, 2004.
13. J. Zeng and W. Qiao, "Support vector machine-based short-term wind power forecasting," in *IEEE/PES Power Systems Conference and Exposition (PSCE)*, pp. 1–8, 2011.
14. E. G. Ortiz-Garcia, S. Salcedo-Sanz, A. M. Perez-Bellido, J. Gascon-Moreno, J. A. Portilla-Figueras, and L. Prieto, "Short-term wind speed prediction in wind farms based on banks of support vector machines," *Wind Energy*, vol. 14, no. 2, pp. 193–207, 2011.
15. S. Santoso, M. Negnevitsky, and N. Hatziargyriou, "Data mining and analysis techniques in wind power system applications: abridged," in *Power Engineering Society General Meeting, 2006. IEEE*, pp. 1–3, 2006.
16. A. Kusiak, H. Zheng, and Z. Song, "Wind farm power prediction: a data-mining approach," *Wind Energy*, vol. 12, no. 3, pp. 275–293, 2009.
17. H. Y. Zheng and A. Kusiak, "Prediction of wind farm power ramp rates: A data-mining approach," *ASME Journal of Solar Energy Engineering*, vol. 131, pp. 031011.1–031011.8, 2009.
18. C. Potter, E. Grimit, and B. Nijssen, "Potential benefits of a dedicated probabilistic rapid ramp event forecast tool," in *Power Systems Conference and Exposition, 2009. PSCE '09. IEEE/PES*, pp. 1–5, 2009.
19. C. Ferreira, J. Gama, L. Matias, A. Botterud, and J. Wang, "A survey on wind power ramp forecasting." Argonne National Laboratory Technical Report, Sept. 2011, available at http://www.dis.anl.gov/pubs/69166.pdf, Apr. 2009.
20. A. Bossavy, R. Girard, and G. Kariniotakis, "Forecasting ramps of wind power production with numerical weather prediction ensembles," *Wind Energy*, vol. 16, pp. 51–63, 2013.
21. H. Zareipour, D. Huang, and W. Rosehart, "Wind power ramp events classification and forecasting: A data mining approach," in *Power and Energy Society General Meeting, 2011 IEEE*, pp. 1–3, 2011.
22. Y. Altun, I. Tsochantaridis, and T. Hofmann, "Hidden markov support vector machines," in *Proceedings of the Twentieth International Conference on Machine Learning (ICML-2003)*, 2003.
23. M. Valstar and M. Pantic, "Combined support vector machines and hidden markov models for modeling facial action temporal dynamics," *Human-computer Interaction, Lecture Notes in Computer Science*, vol. 4796, pp. 118–127, 2007.
24. A. Sloin and D. Burshtein, "Support vector machine training for improved hidden markov modeling," *IEEE Trans. on Signal Process.*, vol. 56, no. 1, pp. 172–188, 2008.
25. "All island grid study—work stream 4, analysis of impacts and benefits." Available at http://www.dcenr.gov.ie/Energy/North-South+Co-operation+in+the+Energy+Sector/All+Island+El ectricity+Grid+Study.htm, January 2008.
26. "WILMAR (Wind Power Integration in Liberalised Electricity Markets)." Available at http://www.wilmar.risoe.dk/index.htm.
27. A. Papavasiliou, S. S. Oren, and R. P. ONeill, "Reserve requirements for wind power integration: A scenario-based stochastic programming framework," *IEEE Trans. Power Syst.*, vol. 26, no. 4, pp. 2197–2206, 2011.
28. A. Khosravi, S. Nahavandi, and D. Creighton, "Prediction intervals for short-term wind farm power generation forecasts," *IEEE Trans. on Sustain. Energy*, vol. 4, no. 3, pp. 602–610, 2013.
29. A. Khosravi and S. Nahavandi, "Combined nonparametric prediction intervals for wind power generation," *IEEE Trans. on Sustain. Energy*, vol. 4, no. 4, pp. 849–856, 2013.
30. C. Wan, Z. Xu, P. Pinson, Z. Y. Dong, and K. P. Wong, "Optimal prediction intervals of wind power generation," *IEEE Trans. on Power Syst.*, vol. 29, no. 3, pp. 1166–1174, 2014.
31. M. He, L. Yang, J. Zhang, and V. Vittal, "A spatio-temporal analysis approach for short-term forecast of wind farm generation," *IEEE Trans. on Power Syst.*, vol. PP, no. 99, pp. 1–12, 2014.
32. P. Pinson, H. A. Nielsen, J. K. Moller, H. Madsen, and G. N. Kariniotakis, "Non-parametric probabilistic forecasts of wind power: required properties and evaluation," *Wind Energy*, vol. 10, no. 6, pp. 497–516, 2007.

33. Q. Zhang and S. A. Kassam, "Finite-state Markov model for Rayleigh fading channels," *IEEE Trans. Commun.*, vol. 47, no. 11, pp. 1688–1692, 1999.
34. C.-W. Hsu and C.-J. Lin, "A comparison of methods for multiclass support vector machines," *IEEE Transactions on Neural Networks*, vol. 13, no. 2, pp. 415–425, 2002.
35. C.-W. Hsu, C.-C. Chang, and C.-J. Lin, "A practical guide to support vector classification," 2010.
36. C.-C. Chang and C.-J. Lin, "LIBSVM: A library for support vector machines," *ACM Transactions on Intelligent Systems and Technology*, vol. 2, pp. 27:1–27:27, 2011. Software available at http://www.csie.ntu.edu.tw/cjlin/libsvm.
37. C. J. Burges, "A tutorial on support vector machines for pattern recognition," *Data Mining and Knowledge Discovery*, vol. 2, pp. 121–167, 1998.
38. M. H. Hayes, *Statistical Digital Signal Processing and Modeling*. New York, NY, USA: Wiley, 1996.
39. T. Gneiting, F. Balabdaoui, and A. E. Raftery, "Probabilistic forecasts, calibration and sharpness," *Journal of the Royal Statistical Society, Series B, Statistical Methodology*, vol. 69, pp. 243–268, 2007.

Chapter 4
Stochastic Optimization Based Economic Dispatch and Interruptible Load Management

In this chapter, stochastic optimization of economic dispatch (ED) and interruptible load management is investigated using short-term distributional forecast of wind farm generation using the proposed short-term distributional forecast of wind farm generation. Based on the distributional forecast model, the joint optimization of ED and interruptible load management is cast as a stochastic optimization problem. Additionally, a robust ED is formulated using an uncertainty set constructed based on the proposed distributional forecast, aiming to minimize the system cost for worst cases. The proposed stochastic ED is compared with three other ED schemes, namely the robust ED, the deterministic ED using the persistence wind generation forecast model, and the genie-aided ED with perfect wind generation forecasts.

4.1 Introduction

In order to meet the renewable portfolio standards (RPS), much effort is being invested to integrate renewable generation (particularly wind generation) into bulk power grids, and wind energy constitutes a significant portion of this renewable integration [1]. High penetration of wind generation, however, is expected to result in significant operational challenges [2], due to its non-dispatchability and variability. Accurate forecasting of future wind generation across temporal and spatial scales still remains elusive [3]. Therefore, the integration of wind generation at high penetration levels into bulk power grids may have significant impact on system reliability, because of the inability to attain an acceptable load/generation balance.

One possible approach to integrate wind generation into power system operational planning is to treat wind generation as negative load [4]. Conventionally, in power system operations, the effects of load forecast errors are mitigated by the regulation and operating reserves that are co-scheduled with generation. However, as pointed out in [5], wind generation is far more variable and unpredictable than the load. It is clear that, at high penetration levels, wind generation would become a dominant factor in terms of uncertainty; and hence there is an urgent need to revisit the approach to scheduling regulation and operating reserves (see, e.g., recent studies [6, 7] on

© The Author(s) 2014
L. Yang et al., *Spatio-Temporal Data Analytics for Wind Energy Integration*,
SpringerBriefs in Electrical and Computer Engineering, DOI 10.1007/978-3-319-12319-6_4

this issue). Further, it is reasonable to expect that the cost of regulation services and operating reserves could significantly increase when the penetration level of wind generation is high.

Aiming to maintain system reliability with high penetration of wind generation and reduce the cost of reserves, joint optimization of economic dispatch (ED) and interruptible load management [8] is explored in this chapter. Interruptible loads have been recognized as one of the ancillary services, particularly, as a contingency reserve service. In an interruptible load program, the customer enters into a contract with the independent system operator (ISO) to reduce its demand when requested. The ISO benefits by reducing its peak load and thereby saving costly reserves, restoring quality of service and ensuring reliability. Consequently, the customer benefits from the reduction in energy costs and from the incentives provided by the contract. By using interruptible load management, the system may conserve costly reserves to accommodate load/generation imbalance due to wind generation forecast errors.

In addition to interruptible load management, accurate wind generation forecast is expected to reduce the requirement of regulation services and operating reserves. A vast amount of existing literature on wind generation forecast focuses on wind speed forecast which is subsequently translated into the wind power output based on the turbine power curve (see, e.g., [9]). In fact, as shown in [10], the power outputs from identical turbines within a farm are not necessarily equal, even if the turbines are co-located. This "mismatch" is more severe when the turbines are located far apart, and could result in an erroneous forecast of wind farm output. Besides, as reported in [11], system operations based on distributional wind generation forecasts can improve the system efficiency. In this work, the Markov-chain-based distributional wind farm generation forecast proposed in Chap. 2 is utilized to improve the system efficiency.

Based on recent work [10, 12], a novel stochastic ED model is exploited by leveraging the Markov-chain-based distributional wind farm generation forecast and interruptible load management, aiming to minimize the system operating costs. In light of the Markovian property of this forecast model, the ED problem is formulated as a stochastic optimization problem that can minimize the expected cost of using operating reserves to compensate for the forecast errors, by accounting for the likelihood of transiting from the current wind generation state to a few neighboring states. It is worth noting that, thanks to the high forecast accuracy of the Markov-chain-based wind farm generation, the system cost can be reduced, as verified by simulations. By using actual wind power data from NREL and Xcel Energy with power outputs of all wind turbines and wind speeds measured at all METs recorded every 10 min from 2008 to 2010, the proposed stochastic ED is tested on the IEEE Reliability Test System—1996 [13]. Such a simulation study is carried out for different wind penetration levels so as to demonstrate economic benefits of the proposed stochastic ED framework.

4.1.1 Related Work

A significant body of work in the literature indicates that operating reserves could be procured from demand response [14], instead of from backup generation capacity. Black et al. [15] propose a methodology for integrating demand response into optimal dispatch algorithms, taking into account the impact of load shifting to later time periods. The work by Papavasiliou et al. [16] presents a contract for integrating renewable energy supply into electricity spot markets for serving deferrable electric loads to mitigate renewable energy intermittency. In [17], three approaches are proposed to integrate short-term responsiveness into a generation technology mix optimization model while considering operational constraints. Recent efforts [18] and [19] use several case studies to demonstrate that real-time pricing could help improve the utilization and cost of integration of wind generation. Unfortunately, a lack of real-time pricing has prevented most consumers from tracking and responding to real-time prices, resulting in an inelastic demand in the short-term [20]. In this paper, interruptible load management is jointly optimized with ED.

In this work, we propose a stochastic ED formulation, based on the Markov-chain-based distributional forecast model [12], which considers the likelihood of transiting from the current wind generation state to a few neighboring states, thus accounting for the range of the wind farm generation in the next time slot. Most existing works on wind power integration utilize either scenario-based stochastic programming or robust optimization to address the uncertainty and variability of wind power (see, e.g., [4, 6, 11, 21–27]). The scenario-based formulations (see, e.g., [4, 6, 11, 21–23]) often ignore the temporal correlation of the wind turbines that would significantly impact wind farm generation, and as a result, the considered scenarios cannot accurately characterize the future wind farm generation. Recently, robust optimization has gained more attention as a modeling framework to account for parameter uncertainty since it provides an effective means to improve system robustness by optimizing the problem under the worst-case scenarios (see, e.g., [24– 27]). Instead of using a probabilistic distribution, robust optimization requires that wind power output falls within a given uncertainty set only and searches for a solution that can ensure system robustness against all realizations within the given uncertainty set. Therefore, the solutions obtained by robust optimization approaches are often considered to be conservative.

Moreover, there is very limited work on the impact of wind power forecast errors on the real-time dispatch in market operation. The forecast errors, which are the "mismatches" between what is scheduled in the ED stage and the actual wind farm output during real-time operation, make it challenging for system operators to balance the unexpected deficit (or surplus) of wind power. Wang et al. [28] investigate the impact of wind power forecasting on unit commitment by using the actual wind generation in ED. In general, even at the ED stage, perfect wind generation forecasts are unattainable due to the variability of wind and thus wind forecast errors also exist in ED. Therefore, one focus of this study is on the impact of wind forecast errors on the real-time dispatch in market operation.

Table 4.1 Summary of the main notation

Notation	Definition
b	Bus index
d_i	Distance from node i to the root of the minimal spanning tree
g	Generator index
w	Wind farm index
n_{ij}	Number of transitions from S_i to S_j
$B_{bb'}$	The susceptance of line bb'
$C_b^I(\cdot)$	Interruptible load cost function at bus b
$C_g^G(\cdot)$	Generation cost function of generator g
$C_w^W(\cdot)$	Cost function of integrating the wind generation of wind farm w
D_b^t	Load at bus b at time t
F_w	Probability distribution of farm aggregate wind generation
L_b^t	Load at bus b that could be interrupted at time t
N_w	Number of states of the Markov chain of wind farm w
N_t	Number of measurement data
$P_w(t)$	Historical aggregate power output of wind farm w
$P_i^w(t)$	Historical power output of wind turbine i in wind farm w
$P_{bb'}$	Branch power flow
$P_{bb'}^{\max}$	Rated capacity of transmission line bb'
P_b^t	Scheduled interruptible load at bus b at time t
P_g^t	Scheduled power output of generator g at time t
$P_g^{\min}, P_g^{\max}$	Minimum/maximum power output of generator g
P_w^t	Power generation from the wind farm w at time t
$P_w^{\min}, P_w^{\max}$	Minimum/maximum generation from wind farm w
$\overline{P}_w^t$	Expectation of wind generation forecast of wind farm w at time t
$\tilde{P}_w^t$	Persistence forecast of the generation from wind farm w at time t
Q	Transition probability matrix
R_g^t	Scheduled regulation reserve from generator g at time t
R_s	Regulation reserve requirement
RP_g^{up}, RP_g^{dn}	Maximum ramping up/down of generation g
S_k	The kth state in $\mathcal{S}, k \in \{1, \cdots, N_w\}$
$\mathcal{B}$	Set of buses
$\mathcal{G}$	Set of generators in the system
$\mathcal{G}_b$	Set of generators at bus b
$\mathcal{N}(\Gamma)$	Level crossing rate for the farm aggregate power Γ

Table 4.1 (continued)

Notation	Definition
$\mathcal{P}_w$	Random process of the aggregate wind farm generation
$\mathcal{S}$	State space of the Markov chain
$\mathcal{W}$	Set of wind farms in the system
$\mathcal{W}_b$	Set of wind farms at bus b
α	Linear regression coefficient for the parent-child turbine pairs
θ_b	Voltage angle of bus b
τ_k	Average duration of state S_k
Γ	Wind farm generation level

The rest of the chapter is organized as follows. In Sect. 4.2, the ED model with interruptible load management is proposed by leveraging the distributional wind power forecast. In Sect. 4.3, an illustrative power system ED example is presented, which quantifies the potential savings in both generation cost and regulation reserves in the proposed ED model. A summary of the chapter is provided in Sect. 4.4. The main notation used in the chapter is summarized in Table 4.1.

4.2 Economic Dispatch with Short-Term Distributional Forecast of Wind Generation

Figure 4.1 illustrates the flow chart of the stochastic economic dispatch (ED) with short-term distributional wind farm generation forecast. The Markov chain based forecast model utilizing the spatio-temporal data analysis is developed in Chap. 2 using the historical wind generation data to extract the statistical features of the wind farm generation during different epochs in an entire year. Then the stochastic ED problem is formulated by integrating the distributional forecast of wind generation and interruptible load management. Worth noting is that load uncertainty and forced outages of generators and transmission lines are not directly considered.

4.2.1 Interruptible Load Management

The forecast errors are inevitable, and result in complications in system operations, since the actual wind generation may be different from the forecast. When the actual wind generation is larger than the forecast, the wind generation may have to be curtailed if sufficient downward reserves from other resources are not present. When the actual wind generation is lower than the forecast, the system needs to compensate for the forecast errors by using regulation reserves (e.g., spinning reserves). Clearly,

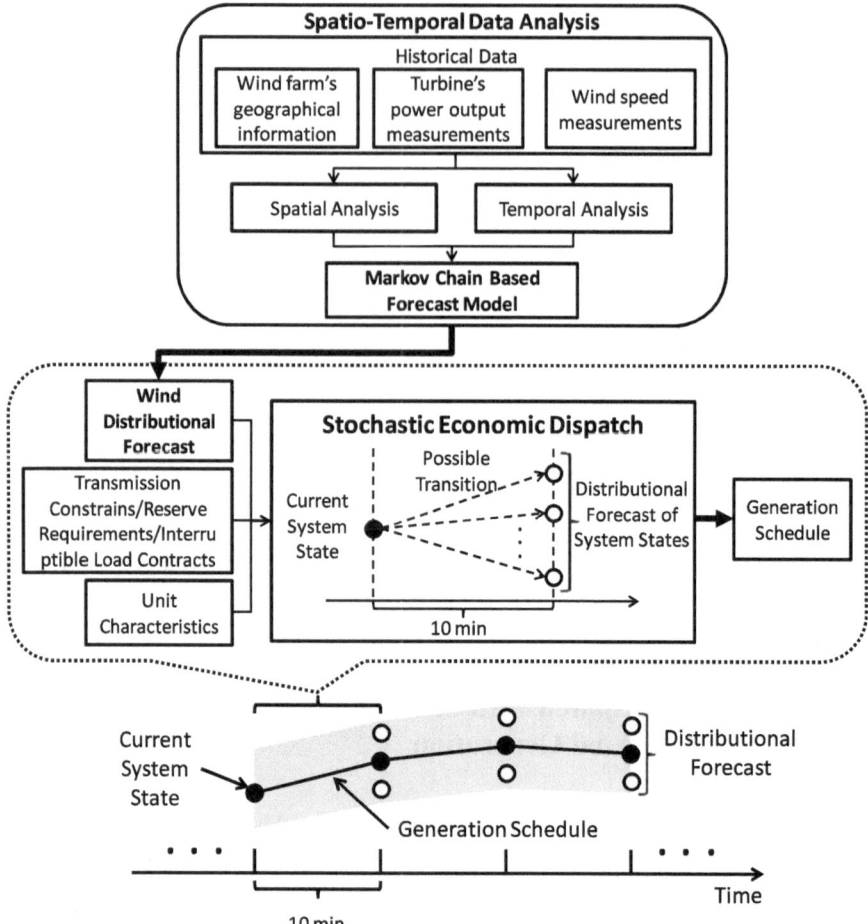

Fig. 4.1 Illustration of stochastic economic dispatch with the short-term distributional forecast

the required reserves increase with the penetration level of wind. Therefore, the reserve cost can be high when the penetration level of wind is high. To reduce the cost, interruptible load services can be used when the actual wind generation is lower than the forecast. The ISO can reduce the demand and thereby withhold costly reserves. In this work, it is assumed that the interruptible load contracts are given a priori, i.e., the amount of load (L_b^t) that can be interrupted in each slot is known [29, 30].

4.2.2   Stochastic Optimization Based Economic Dispatch

In this section, the stochastic ED problem is formulated by using both spatio-temporal wind generation forecast and interruptible load management. Specifically, leveraging the Markov-chain-based forecast model, the ED problem is cast as a stochastic optimization problem (**P1**). For the purpose of performance comparison, a robust ED (**P2**) is formulated to minimize the system operation costs for the worst cases under an uncertainty set constructed based on the forecast distribution given by the Markov-chain-based forecast model. Furthermore, a deterministic ED problem using the persistence forecast of wind farm generation (**P3**) is also formulated as a benchmark.

4.2.2.1 Stochastic ED (P1)

P1 minimizes the system operation costs including costs of generation ($C_g^G(P_g^t)$), costs of using interruptible load services ($C_b^I(P_b^t)$), and costs of integrating wind generation ($\mathbb{E}\{C_w^W(P_w^t)\}$) that take into account the expected costs of using reserves to compensate for the forecast errors when the actual wind generation is lower than the forecast.

$$\textbf{P1}: \text{minimize} \sum_{g \in \mathcal{G}} C_g^G(P_g^t) + \sum_{b \in \mathcal{B}} C_b^I(P_b^t) + \sum_{w \in \mathcal{W}} \mathbb{E}\{C_w^W(P_w^t)\} \tag{4.1}$$

$$\text{subject to} \sum_{g \in \mathcal{G}_b} P_g^t + \sum_{w \in \mathcal{W}_b} P_w^t - D_b^t + P_b^t = \sum_{b' \in \mathcal{B}, b \neq b'} P_{bb'}, \forall b \in \mathcal{B} \tag{4.2}$$

$$P_{bb'} = B_{bb'}(\theta_b' - \theta_b), \forall b, b' \in \mathcal{B}, b \neq b' \tag{4.3}$$

$$|P_{bb'}| \leq P_{bb'}^{\max}, \forall b, b' \in \mathcal{B}, b \neq b' \tag{4.4}$$

$$\sum_{g \in \mathcal{G}} R_g^t \geq R_s \tag{4.5}$$

$$-RP_g^{dn} \leq P_g^t - P_g^{t-1} \leq RP_g^{up}, \forall g \in \mathcal{G} \tag{4.6}$$

$$P_g^{\min} \leq P_g^t + R_g^t \leq P_g^{\max}, \forall g \in \mathcal{G} \tag{4.7}$$

$$0 \leq P_b^t \leq L_b^t, \forall b \in \mathcal{B} \tag{4.8}$$

$$P_w^{\min} \leq P_w^t \leq P_w^{\max}, \forall w \in \mathcal{W} \tag{4.9}$$

where

- In (4.1), with the proposed distributional forecast method, we have $\mathbb{E}\{C_w^W(P_w^t)\} = \sum_{j \in \mathcal{S}} Q_{S_w^{t-1}, j} c_R (P_w^t - P_{w,j})^+$, where $Q_{S_w^{t-1}, j}$ is given by the distributional wind generation forecast (i.e., (2.13) in Chap. 2), S_w^{t-1} denotes the current wind generation state of wind farm w, c_R is the per unit cost of reserve, and $(x)^+ = \max(x, 0)$. The quadratic production cost function is used, i.e., $C_g^G(P_g^t) = a_g + b_g P_g^t + c_g (P_g^t)^2$, and the cost of using interruptible load services is assumed to be linear, i.e., $C_b^I(P_b^t) = c_I P_b^t$, where c_I is the per unit cost of using interruptible load services.
- (4.2) is the power balance at bus b.

- In (4.3), $P_{bb'}$ is the real power injection from bus b to bus b' based on the DC power flow approximation, where $B_{bb'}$ is the susceptance of line bb', and θ_b and θ'_b are voltage angles of bus b and bus b', respectively.
- (4.4) is the rated capacity of transmission line bb'.
- (4.5) is the system reserve requirement, which can be determined by the system reliability requirement, the wind forecast accuracy and load forecast accuracy.
- (4.6) is the ramping constraint of each conventional generator.
- (4.7) is the capacity constraint of each generator for providing reserve services.
- (4.8) is the interruptible load constraint specified by the interruptible load contracts, where L_b^t denotes the amount of load that can be interrupted at bus b.
- (4.9) is the upper and lower bound of wind farm output.

When solving **P1**, $\mathbb{E}\{C_w^W(P_w^t)\}$ is replaced by auxiliary variables and a set of linear constraints. Specifically, each component $Q_{S_w^{t-1},j}c_R(P_w^t - P_{w,j})^+$ in $\mathbb{E}\{C_w^W(P_w^t)\}$ is replaced by an auxiliary variable $P_{w,j}^t$ and the constraints $P_{w,j}^t \geq Q_{S_w^{t-1},j}c_R(P_w^t - P_{w,j})$ and $P_{w,j}^t \geq 0$. By this transformation, **P1** can be solved by many commercial solvers, e.g., CPLEX [31]. Note that different from the scenario-based formulations that more possible wind generation states would be considered, **P1** considers only a few neighboring states of the current wind generation state by using the proposed distributional forecast (e.g., as illustrated in Fig. 2.11a, the transition probabilities from the current state to most non-adjacent states are zero). This would not only reduce the computational complexity since fewer states are considered, but also improve the system operation cost since our distributional forecast is more accurate due to the consideration of the temporal correlation of wind generation, as shown in Sect. 4.3.

4.2.2.2 Robust ED (P2)

The robust ED minimizes system operation costs for worst cases under an uncertainty set. The uncertainty set $\mathcal{U}$ is constructed based on the forecast distribution given by the Markov-chain-based forecast model:

$$\mathcal{U}(\overline{\mathbf{P}}_w^t, \mathbf{LB}_w^t, \mathbf{UB}_w^t, \Gamma^t) = \{\hat{\mathbf{P}}_w^t \in \mathbb{R}^{|\mathcal{W}|} : \sum_{w \in \mathcal{W}} \frac{\hat{P}_w^t - \overline{P}_w^t}{UB_w^t - \overline{P}_w^t} \leq \Gamma^t,$$

$$\sum_{w \in \mathcal{W}} \frac{\overline{P}_w^t - \hat{P}_w^t}{\overline{P}_w^t - LB_w^t} \leq \Gamma^t, \ \hat{P}_w^t \in [LB_w^t, UB_w^t], \forall w \in \mathcal{W}\}.$$

$$(4.10)$$

Here $\hat{\mathbf{P}}_w^t$ is the vector of wind generation forecasts. $\overline{\mathbf{P}}_w^t$ is the vector of expectations of wind generation forecast at time t given by the Markov-chain-based forecast model. $\mathbf{LB}_w^t$ and $\mathbf{UB}_w^t$ are the lower bounds and upper bounds of wind forecast. Γ^t is defined as the budget of uncertainty for wind forecast, which takes the value between 0 and $|\mathcal{W}|$, where $|\mathcal{W}|$ is the number of wind farms. When $\Gamma^t = 0$, the uncertainty set $\mathcal{U}$ is a singleton, corresponding to the deterministic ED formulation. As Γ^t increases, the size of the uncertainty set $\mathcal{U}$ enlarges, indicating that the resulting solutions are more conservative and the system is protected against a higher degree of uncertainty.

Using $\mathcal{U}$, the robust ED is formulated as

$$\mathbf{P2}:\quad \text{minimize}\quad \sum_{g\in G} C_g^G(P_g^t) + \sum_{b\in B} C_b^I(P_b^t) + \max_{\hat{\mathbf{P}}_w^t \in \mathcal{U}} \sum_{w\in W} c_R(P_w^t - \hat{P}_w^t)^+$$

$$\text{subject to}\quad (4.2),(4.3),(4.4),(4.5),(4.6),(4.7),(4.8),(4.9).$$

$$(4.11)$$

When solving $\mathbf{P2}$, $\max_{\hat{\mathbf{P}}_w^t \in \mathcal{U}} \sum_{w\in W} c_R(P_w^t - \hat{P}_w^t)^+$ can be replaced by auxiliary variables and a set of linear constraints. $\mathbf{P2}$ can be rewritten as

$$\mathbf{P2}:\quad \text{minimize}\quad \sum_{g\in G} C_g^G(P_g^t) + \sum_{b\in B} C_b^I(P_b^t) + C^W$$

$$\text{subject to}\quad (4.2),(4.3),(4.4),(4.5),(4.6),(4.7),(4.8),(4.9)$$

$$C^W \geq \sum_{w\in W} c_w^W,\ c_w^W \geq 0, \forall w \in \mathcal{W}$$

$$c_w^W \geq c_R(P_w^t - \hat{P}_w^t), \forall \hat{\mathbf{P}}_w^t \in \mathcal{U}, \forall w \in \mathcal{W}.$$

$$(4.12)$$

4.2.2.3   Deterministic ED (P3)

For comparing the performance of the stochastic ED (**P1**) and the robust ED (**P2**), the deterministic ED (**P3**) based on the persistence forecast of wind farm generation is formulated as a benchmark:

$$\mathbf{P3}:\quad \text{minimize}\quad \sum_{g\in G} C_g^G(P_g^t) + \sum_{b\in B} C_b^I(P_b^t) + \sum_{w\in W} c_R(P_w^t - \tilde{P}_w^t)^+$$

$$\text{subject to}\quad (4.2),(4.3),(4.4),(4.5),(4.6),(4.7),(4.8),(4.9),$$

$$(4.13)$$

where $\tilde{P}_w^t$ denotes the persistence forecast of wind farm generation that equals to the actual wind farm generation in the previous time slot.

Remarks:

- Both **P1** and **P2** consider the uncertainty of wind generation by using the distributional forecast of wind farm generation. **P2** utilizes only the expectation and the upper and lower bounds of the distributional forecast to minimize the system costs for worst cases, whereas **P1** uses all the information of the distributional forecast. Intuitively speaking, **P2** could result in a more conservative solution in terms of wind generation schedule than **P1**, as illustrated in Sect. 4.3 based on realistic wind measurement data.
- Compared to **P1** and **P2**, the deterministic ED solution by **P3** does not consider the uncertainty of wind generation by using the persistence forecast of wind farm generation, and may schedule the wind generation as the forecasted wind generation. Therefore, **P3** could result in a higher system operation cost, since more reserves may be used to compensate for the forecast errors when the actual

wind generation is less than the forecast, as illustrated in Sect. 4.3 based on realistic wind measurement data.

4.3 Case Studies

4.3.1 Data and Simulation Setup

In this section, the proposed stochastic ED formulation is applied to the IEEE Reliability Test System (RTS)—1996 [13] under different wind penetration levels to simulate the impact of using different wind generation forecasts on the system. In the RTS-1996 system, the large coal-fired plant U350 in each area is assumed to be replaced by a wind farm. The wind generation data from 2009 to 2010 from NREL and Xcel Energy is used in the study, where the wind generation data in 2009 is used to develop the Markov-chain-based distributional forecast model, and the wind generation data in 2010 is used to evaluate the proposed stochastic ED formulation, after proper scaling to suit the chosen penetration level.

The simulation duration is 1 year, and in the case studies, the wind generation data in 2010 is used. For ease of exposition, a deterministic UC is solved for the 24-h period using day-ahead wind power forecasts based on the day-ahead wind power predictor [22], and then the ED problem is solved for every 10 min based on the results of the UC. In UC, typical technical restrictions are accounted for, such as the minimum up and down time limits and the startup costs of generators [32]. Specifically, a cold startup cost and a warm startup cost are considered, depending on the length of time that the unit is down. Then, ED is run every 10 min for 24 h. In both UC and ED, the "3 + 5" rule is used to determine the regulation reserve requirement R_s, which equals 3 % of hourly forecast load plus 5 % of hourly forecast wind power [33]. UC and ED are run in sequence for the entire year to evaluate the proposed stochastic ED formulation.

The hourly profile of the loads is taken from Table 4 of [13] with a peak value 8550 MW. The characteristics of the power plants are obtained from Table 10 of [13], where the startup costs of the power plants are computed based on the fuel cost taken from Table II of [34]. The quadratic generation cost functions defined in Table 1 of [35] are used here. The coefficient of the interruptible load cost function c_I and the reserve cost function c_R are \$ 10/MWh and \$ 5000/MWh [24], respectively.

In the simulations, the proposed stochastic ED (**P1**) is run to compare with the solutions of the robust ED (**P2**) and that of the deterministic ED (**P3**) using the persistence forecast of wind farm generation. Besides, the solutions of *the genie-aided ED* with perfect wind generation forecasts are also provided as a benchmark, in which the actual wind farm generation is used as the forecast. Specifically, in the genie-aided ED, the objective is to minimize the costs of generation $(C_g^G(P_g^t))$ and the costs of using interruptible load services $(C_b^I(P_b^t))$ subject to the same constraints in **P3**. Obviously, the system cost of the genie-aided ED based on the perfect forecast

Table 4.2 System cost with 10 % wind penetration

Month	System cost[a]			
	P1	**P2**	**P3**	Genie-aided ED
January	4.6498	4.6912	4.8576	4.5724
February	4.1761	4.2167	4.3766	4.1090
March	4.6182	4.6672	4.8262	4.5404
April	4.3468	4.3998	4.5860	4.2671
May	4.5973	4.6530	4.8090	4.5031
June	4.5690	4.6124	4.8149	4.4809
July	4.9594	4.9911	5.1977	4.8745
August	4.7646	4.7980	5.0156	4.6819
September	4.4103	4.4501	4.6299	4.3393
October	4.5851	4.6292	4.8005	4.5180
November	4.4265	4.4755	4.6662	4.3490
December	4.5519	4.6012	4.7720	4.4750L
Total	54.655	55.185	57.352	53.711
Total improv. (%)		0.97	4.93	−1.73

[a] The numbers in these tables should be multiplied by 10^8 to obtain the cost in $

is the lowest among these formulations, as there are no forecast errors and thereby no reserves are consumed to compensate for the forecast errors.

In the simulations, planned and forced outages of generators are not considered, since the focus of this work is to investigate the impact of wind generation on system operation. The ED problems are solved by using CPLEX 12.5 [31] on a PC with a 2.4 GHz Intel Core i3 processor and 4 GB RAM. Note that, since the forecast errors are accounted for, i.e., the forecasted wind generation may be different from the actual wind generation, the real system cost is computed based on the actual wind generation.

4.3.2 Results and Discussions

In this section, the results of the real system costs given by **P1**, **P2**, **P3**, and the genie-aided ED are provided to illustrate the impact of different short-term wind generation forecasts and the problem formulations. In **P2**, Γ^t is chosen to be $|\mathcal{W}|$ in the simulations.

Tables 4.2, 4.3, and 4.4 compare the system costs of **P1**, **P2**, **P3**, and the genie-aided ED under different wind penetration levels. As discussed in Chap. 2, the Markov chain is non-stationary, due to the diurnal non-stationarity and the seasonality of wind speed. In the simulations, the Markov-chain-based forecast model is designed for

Table 4.3 System cost with 20 % wind penetration

Month	System cost[a]			
	P1	P2	P3	Genie-aided ED
January	4.3279	4.3831	4.7785	4.2112
February	3.8594	3.9140	4.2941	3.7606
March	4.2456	4.3107	4.6928	4.1312
April	3.9486	4.0167	4.4644	3.8352
May	4.1616	4.2355	4.6340	4.0288
June	4.2370	4.2953	4.7662	4.1017
July	4.6801	4.7279	5.1854	4.5439
August	4.4876	4.5377	5.0173	4.3568
September	4.1090	4.1630	4.5765	4.0002
October	4.2658	4.3209	4.7253	4.1638
November	4.0655	4.1270	4.5791	3.9537
December	4.1954	4.2550	4.6746	4.0871
Total	50.583	51.287	56.388	49.174
Total improv. (%)		1.39	11.48	−2.79

[a] The numbers in these tables should be multiplied by 10^8 to obtain the cost in $

Table 4.4 System cost with 30 % wind penetration

Month	System cost[a]			
	P1	P2	P3	Genie-aided ED
January	4.1011	4.1644	4.8008	3.9583
February	3.6372	3.6976	4.3107	3.5199
March	3.9846	4.0552	4.6846	3.8472
April	3.6724	3.7462	4.4616	3.5358
May	3.8623	3.9418	4.5957	3.7066
June	3.9975	4.0635	4.7872	3.8321
July	4.4865	4.5432	5.2658	4.3152
August	4.2756	4.3354	5.0894	4.1140
September	3.8880	3.9482	4.6053	3.7566
October	4.0397	4.1029	4.7444	3.9138
November	3.8091	3.8778	4.5931	3.6762
December	3.9487	4.0159	4.6818	3.8147
Total	47.703	48.492	56.620	45.990
Total improv. (%)		1.65	18.69	−3.59

[a] The numbers in these tables should be multiplied by 10^8 to obtain the cost in $

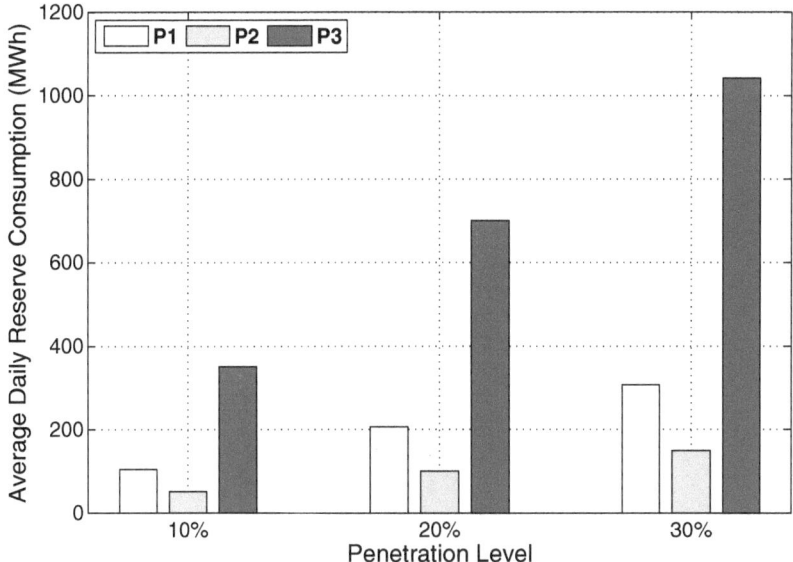

Fig. 4.2 Average daily reserve consumption under different formulations at different penetration levels

each epoch in each month, using the historical data. The system costs of each month are illustrated in Tables 4.2, 4.3, and 4.4. The row with total costs sums the cost of each month. The last row shows the total improvement of **P1** over **P2**, **P3**, and the genie-aided ED, normalized by the total cost of **P1**, where the negative value means that the total system cost of **P1** is higher than that of the genie-aided ED.

It is observed from Tables 4.2, 4.3, and 4.4 that the system cost in each month (and the yearly total system cost) is indeed improved using **P1**, compared with **P2** and **P3**, and the improvement increases with the wind penetration level. This shows that (1) the consideration of the uncertainty of wind generation in ED can indeed help to reduce the system cost and (2) the proposed distributional forecast is accurate, as the system cost of **P1** is less than that of **P2**, and close to the genie-aided ED. This demonstrates the benefits of the proposed stochastic ED (**P1**) using the proposed Markov-chain-based forecast model and interruptible load management.

Figure 4.2 compares the average daily reserve needed to compensate for the forecast errors when the actual wind generation is less than the scheduled wind generation. As the genie-aided ED uses the perfect wind generation forecast, no reserve is consumed. As shown in Fig. 4.2, the average daily reserve consumed in **P1** is much less than that in **P3**, which explains why the system cost of **P1** is much less than that of **P3**. From Fig. 4.2, we also observe that the average daily reserve consumed in **P2** is the least, which is due to the fact that **P2** minimizes the system costs for worst cases and thereby results in the most conservative wind generation schedule.

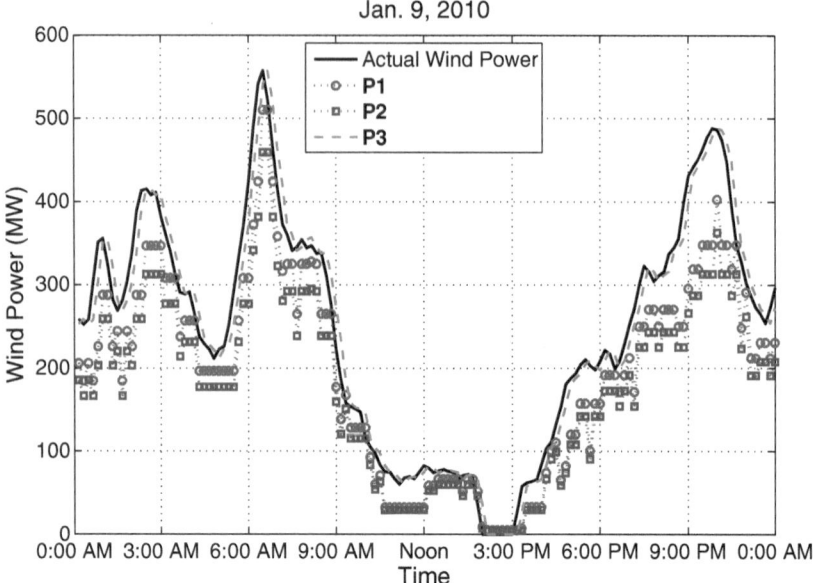

Fig. 4.3 Comparison of wind generation schedule of **P1**, **P2**, and **P3**, when the wind penetration level is 30 %

Figure 4.3 compares the wind generation schedule of one wind farm given by **P1**, **P2**, and **P3** on Jan. 9th in 2010. As the genie-aided ED uses the perfect wind generation forecast, the wind generation schedule of the genie-aided ED corresponds to the actual wind generation. In Fig. 4.3, the solid curve denotes the actual wind generation, and the curves with circle and square, and the dash curve denote the wind generation schedule given by **P1**, **P2**, and **P3**, respectively. It is observed that at most times, the wind generation schedules of **P1** and **P2** are below the actual wind generation, and thereby **P1** and **P2** can significantly save the costly reserves used to compensate for the forecast errors when the actual wind generation is less than the wind generation schedule. Moreover, the wind generation schedule of **P2** is more conservative than that of **P1**, in the sense that less wind generation is scheduled. Therefore, compared with **P1**, **P2** would schedule more conventional generation and thereby result in a higher system cost, as shown in Tables 4.2, 4.3, and 4.4. On the contrary, **P3** ignores the uncertainty of the wind generation and tries to utilize the wind generation based on the persistence forecast, which may result in the consumption of more reserves when the actual wind generation ramps down as illustrated in Fig. 4.3, where the scheduled wind generation of **P3** is above the actual wind generation at most times. Therefore, although **P3** can utilize more wind generation, **P3** would result in a much higher system cost due to the costly reserves, as shown in Tables 4.2, 4.3, and 4.4. Compared with **P2** and **P3**, **P1** strikes a tradeoff between the utilization of the wind generation and the possible reserves needed to compensate for the forecast errors.

Simply put, the benefits of the proposed stochastic ED (**P1**) result from the proposed Markov-chain-based distributional forecast model and interruptible load management. Specifically, **P1** takes into account the uncertainty of the wind generation and schedules the wind generation more conservatively than **P3**. By scheduling the wind generation conservatively, **P1** can reduce the reserves needed to compensate for the forecast errors when the actual wind generation is less than the scheduled wind generation. Besides, as the proposed Markov-chain-based distributional forecast model is accurate, using all the distributional forecast information, **P1** can utilize more wind generation and schedule conventional generation more efficiently than **P2**, in which partial forecast information is utilized.

4.4 Summary

In this chapter, joint stochastic optimization of ED and interruptible load management was investigated using short-term wind farm generation forecast. Specifically, using the finite state Markov chain model for wind farm generation developed in Chap. 2, the joint optimization of ED and interruptible load management was cast as a stochastic optimization problem that can minimize the expected cost of using operating reserves to compensate for the forecast errors, by considering a few neighboring states of the current wind generation state. Numerical studies, via the IEEE Reliability Test System—1996 and realistic wind measurement data from an actual wind farm, demonstrated the significant benefits obtained by integrating the Markov-chain-based distributional forecast and the interruptible load management, compared with the robust ED, the deterministic ED with the persistence forecasts, and the genie-aided ED with the perfect forecasts. In future work, the development of systematic approaches to address the uncertainty of wind generation in electricity markets in a cost-effective manner will be considered. For example, secondary ancillary services markets can be considered in the proposed ED framework.

References

1. K. Cory and B. Swezey, "Renewable portfolio standards in the states: balancing goals and implementation strategies." NREL Technical Report TP-670-41409, Dec. 2007.
2. L. Xie, P. Carvalho, L. Ferreira, J. Liu, B. Krogh, N. Popli, and M. Ilic, "Wind integration in power systems: operational challenges and possible solutions," *Proc. IEEE*, vol. 99, no. 1, pp. 214–232, 2011.
3. K. Porter and J. Rogers, "Survey of variable generation forecasting in the west." NREL Subcontract Report SR-5500–54457, Apr. 2012.
4. F. Bouffard and F. Galiana, "Stochastic security for operations planning with significant wind power generation," *IEEE Trans. Power Syst.*, vol. 23, no. 2, pp. 306–316, 2008.
5. "Accommodating high levels of variable generation." NERC Special Report, www.nerc.com/files/IVGTF_Report_041609.pdf, Apr. 2009.

6. J. Morales, A. Conejo, and J. Perez-Ruiz, "Economic valuation of reserves in power systems with high penetration of wind power," *IEEE Trans. Power Syst.*, vol. 24, no. 2, pp. 900–910, 2009.
7. M. Ortega-Vazquez and D. Kirschen, "Estimating the spinning reserve requirements in systems with significant wind power generation penetration," *IEEE Trans. Power Syst.*, vol. 24, no. 1, pp. 114–124, 2009.
8. K. Bhattacharya, M. Bollen, and J. Daalder, *Operation of Restructured Power Systems*. Kluwer Academic Publishers, London, 2001.
9. G. Giebel, R. Brownsword, and G. Kariniotakis, "The state of the art in short-term prediction of wind power: A literature overview, 2nd edition," *Project report for the Anemos.plus and SafeWind projects*, 2011.
10. S. Murugesan, J. Zhang, and V. Vittal, "Finite state Markov chain model for wind generation forecast: A data-driven spatio-temporal approach," in *Innovative Smart Grid Technologies, IEEE PES*, pp. 1–8, Jan. 2012.
11. A. Papavasiliou, S. S. Oren, and R. P. ONeill, "Reserve requirements for wind power integration: A scenario-based stochastic programming framework," *IEEE Trans. Power Syst.*, vol. 26, no. 4, pp. 2197–2206, 2011.
12. M. He, L. Yang, J. Zhang, and V. Vittal, "A spatio-temporal analysis approach for short-term wind-farm power generation forecast," *IEEE Trans. Power Syst.*, vol. 29, no. 4, pp. 1611–1622, 2014.
13. C. Grigg *et al.*, "The IEEE reliability test system-1996," *IEEE Trans. Power Syst.*, vol. 14, no. 3, pp. 1010–1020, 1999.
14. B. J. Kirby, "Spinning reserve from responsive loads." Oak Ridge National Laboratory, Mar. 2003.
15. J. Black, J. de Bedout, and R. Tyagi, "Incorporating demand resources into optimal dispatch," in *Energy 2030 Conference, 2008. ENERGY 2008. IEEE*, pp. 1–8, Nov. 2008.
16. A. Papavasiliou and S. S. Oren, "Integration of contracted renewable energy and spot market supply to serve flexible loads," in *18th World Congress of the International Federation of Automatic Control*, pp. 1–8, Aug. 2011.
17. C. De Jonghe, B. Hobbs, and R. Belmans, "Optimal generation mix with short-term demand response and wind penetration," *IEEE Trans. Power Syst.*, vol. 27, no. 2, pp. 830–839, 2012.
18. R. Sioshansi and W. Short, "Evaluating the impacts of real-time pricing on the usage of wind generation," *IEEE Trans. Power Syst.*, vol. 24, no. 2, pp. 516–524, 2009.
19. R. Sioshansi, "Evaluating the impacts of real-time pricing on the cost and value of wind generation," *IEEE Trans. Power Syst.*, vol. 25, no. 2, pp. 741–748, 2010.
20. G. Strbac, "Demand-side view of electricity markets," *IEEE Trans. Power Syst.*, vol. 18, no. 2, pp. 520–527, 2003.
21. P. A. Ruiz, C. R. Philbrick, and P. W. Sauer, "Wind power day-ahead uncertainty management through stochastic unit commitment policies," in *Power Systems Conference and Exposition*, pp. 1–9, March 2009.
22. J. F. Restrepo and F. D. Galiana, "Assessing the yearly impact of wind power through a new hybrid deterministic/stochastic unit commitment," *IEEE Trans. Power Syst.*, vol. 26, no. 1, pp. 401–410, 2011.
23. E. M. Constantinescu, V. M. Zavala, M. Rocklin, S. Lee, and M. Anitescu, "A computational framework for uncertainty quantification and stochastic optimization in unit commitment with wind power generation," *IEEE Trans. Power Syst.*, vol. 26, no. 1, pp. 431–441, 2011.
24. D. Bertsimas, E. Litvinov, X. A. Sun, J. Zhao, and T. Zheng, "Adaptive robust optimization for the security constrained unit commitment problem," *IEEE Trans. Power Syst.*, vol. 28, no. 1, pp. 52–63, 2013.
25. C. Zhao and Y. Guan, "Unified stochastic and robust unit commitment," *IEEE Trans. Power Syst.*, vol. 28, pp. 3353–3361, Aug. 2013.
26. R. Jiang, J. Wang, M. Zhang, and Y. Guan, "Two-stage minimax regret robust unit commitment," *IEEE Trans. Power Syst.*, vol. 28, pp. 2271–2282, Aug. 2013.

27. L. Xie, Y. Gu, X. Zhu, and M. Genton, "Short-term spatio-temporal wind power forecast in robust look-ahead power system dispatch," *IEEE Trans. Smart Grid*, vol. 5, no. 1, pp. 511–520, 2014.

28. J. Wang, A. Botterud, V. Miranda, C. Monteiro, and G. Sheble, "Impact of wind power forecasting on unit commitment and dispatch," in *Proc. 8th Int. Workshop Large-Scale Integration of Wind Power into Power Systems*, pp. 1–8, Oct. 2009.

29. "(1998) voluntary load curtailment program. power pool of Alberta.," *Available online:http://www.powerpool.ab.ca*.

30. C. S. Chen and J. T. Leu, "Interruptible load control for Taiwan power company," *IEEE Trans. Power Syst.*, vol. 5, no. 2, pp. 460–465, 1990.

31. IBM ILOG, "Introducing ibm ilog cplex optimization studio v12.5.1." http://pic.dhe.ibm.com/infocenter/cosinfoc/v12r5/index.jsp.

32. M. Carrion and J. M. Arroyo, "A computationally efficient mixed integer linear formulation for the thermal unit commitment problem," *IEEE Trans. Power Syst.*, vol. 21, no. 3, pp. 1371–1378, 2006.

33. "Western wind and solar integration study," *National Renewable Energy Laboratory, Tech. Rep.*, May 2010.

34. K. Hedman, M. Ferris, R. O'Neill, E. Fisher, and S. Oren, "Co-optimization of generation unit commitment and transmission switching with N-1 reliability," *IEEE Trans. Power Syst.*, vol. 25, no. 2, pp. 1052–1063, 2010.

35. Georgia Tech Power Systems Control and Automation Laboratory. Avaialble online:http://pscal.ece.gatech.edu/testsys/index.html.

Chapter 5
Conclusions and Future Works

In this chapter, we summarize the main results presented in this monograph and highlight future research directions.

5.1 Conclusions

In this monograph, we have investigated spatio-temporal data analytics for wind energy integration. We conclude the brief with the following remarks.

- During the last decade, wind power has been the fastest in growth among all renewable energy resources. With a prospective high penetration level, wind generation integration is expected to change dramatically the existing operating practices (e.g., unit commitment, economic dispatch and ancillary services procurement) that are critical to the adequacy of bulk power systems. Compared to conventional generation, it is very challenging to integrate wind generation due to the non-dispatchability, variability and uncertainty of wind generation. To efficiently integrate wind power into power systems, accurate wind power forecasting algorithms across multiple time scales (from minutes to weeks) need to be developed, depending on the system operations.
- Short-term wind power forecast is critical for optimization of the scheduling of conventional power plants, particularly for economic dispatch and unit commitment. Although many wind power forecast approaches have been developed, spatio-temporal dynamics of wind generation have received little attention. One key observation of this study is the wind farm spatial dynamics, i.e., the power outputs of wind turbines within the same wind farm can be quite different, even if the wind turbines are of the same class and physically located close to each other. Based on data analytics of historical wind generation data, finite-state Makrov models are developed to model such spatio-temporal dyanmics of wind generation.
- Wind generation forecast errors could result in either committing more conventional generation capacity than needed when the actual wind generation is above the forecast value, or using costly ancillary services and fast acting reserves when

© The Author(s) 2014   77
L. Yang et al., *Spatio-Temporal Data Analytics for Wind Energy Integration,*
SpringerBriefs in Electrical and Computer Engineering, DOI 10.1007/978-3-319-12319-6_5

the actual wind generation is less than the forecast value. The latter situation becomes more significant in the presence of wind ramps, as wind ramps introduce significant uncertainty in wind generation. Therefore, it is imperative to develop accurate forecast approaches for wind farm generation, especially for wind power ramps. Along this direction, we enhance the developed finite-state Makrov models by incorporating wind ramp forecasting based on SVM.

- Based on the developed distributional forecast model, the joint optimization of ED and interruptible load management is cast as a stochastic optimization problem. Most existing works on wind power integration utilize either scenario-based stochastic programming or robust optimization to address the uncertainty and variability of wind power. Compared with these works, the proposed stochastic ED formulation considers the likelihood of transiting from the current wind generation state to a few neighboring states, thus accounting for the range of the wind farm generation in the next time slot.

5.2 Future Research Directions

In this brief, we have investigated spatio-temporal data analytics for wind energy integration. Our solutions focus on short-term wind power forecast for a single wind farm. The efficient integration of wind energy into power systems requires wind power forecasting systems for different time scales. We close this chapter and the monograph with the following future research directions in this field.

- Spatio-Temporal Analysis of Power Outputs across Wind Farms: As shown by our initial studies of massive power output data, wind generation tends to be strongly correlated across multiple wind farms within a geographical region. Traditional approaches of wind power forecast focus on the forecast for each individual wind farm, which ignores the strong correlation of wind farm generation within a region. Therefore, it is of great interest to study the wind power forecast by leveraging such correlation structure.
- Markov Model based Stochastic Optimization: Scenario-based stochastic optimization approaches [1–11] have been studied to manage the uncertainty of wind generation, where pre-sampling discrete scenarios are generated based on a certain probabilistic distribution. A large number of scenarios are needed to obtain a solution with reasonable accuracy, which is computationally intensive. Recently, robust optimization has gained popularity as a modeling framework under parameter uncertainty since it provides an effective means to improve system robustness by optimizing the problem under the worst-case scenarios [12–18]. Instead of using a probabilistic distribution, robust optimization requires wind power output within a given uncertainty set only and searches for a solution that can ensure system robustness against all realizations within the given uncertainty set. Nevertheless, the solutions obtained by robust optimization approaches are often considered to be conservative. Markov model based stochastic optimization

can overcome the above difficulties based on state transition matrices instead of scenarios to integrate intermittent and uncertain wind generation, and therefore deserves further study.

- Characterization of Wind Power Ramps: The ability to precisely control generation power output to meet load forecasts and mitigate emergencies is critical in a power system. The high penetration of wind energy generation, with their non-stationary and variable generation characteristics, however, introduces difficult-to-control dynamics and challenges for power system operation. Wind power ramps in wind farms, rare but high-impact events, make power system operation even more challenging. An accurate characterization of wind ramps is important in determining the reserve requirements needed to maintain a reliable power system in scenarios of high wind penetration. Since large wind power ramps are "rare events," there could be only a few data samples for modeling such extreme events. Therefore, it is of great imp
ortance to develop computational approaches for characterizing such extreme events with a few data samples.

References

1. D. Chattopadhyay and R. Baldick, "Unit commitment with probabilistic reserve," in *Proc. IEEE PES Winter Meeting*, vol. 1, pp. 280–285, 2002.
2. U. Ozturk, M. Mazumdar, and B. Norman, "A solution to the stochastic unit commitment problem using chance constrained programming," *IEEE Trans. Power Syst.*, vol. 19, pp. 1589–1598, Aug. 2004.
3. F. Bouffard, F. Galiana, and A. Conejo, "Market-clearing with stochastic security-Part I: Formulation," *IEEE Trans. Power Syst.*, vol. 20, pp. 1818–1826, Nov. 2005.
4. F. Bouffard, F. Galiana, and A. Conejo, "Market-clearing with stochastic security-Part II: Case studies," *IEEE Trans. Power Syst.*, vol. 20, pp. 1827–1835, Nov. 2005.
5. M. Ortega-Vazquez and D. Kirschen, "Optimizing the spinning reserve requirements using a cost/benefit analysis," *IEEE Trans. Power Syst.*, vol. 22, pp. 24–33, Feb. 2007.
6. L. Wu, M. Shahidehpour, and T. Li, "Stochastic security-constrained unit commitment," *IEEE Trans. Power Syst.*, vol. 22, pp. 800–811, May 2007.
7. J. Wang, M. Shahidehpour, and Z. Li, "Security-constrained unit commitment with volatile wind power generation," *IEEE Trans. Power Syst.*, vol. 23, pp. 1319–1327, Aug. 2008.
8. L. Wu, M. Shahidehpour, and T. Li, "Cost of reliability analysis based on stochastic unit commitment," *IEEE Trans. Power Syst.*, vol. 23, pp. 1364–1374, Aug. 2008.
9. A. Papavasiliou and S. S. Oren, "A stochastic unit commitment model for integrating renewable supply and demand response," in *IEEE PES General Meeting*, pp. 1–6, IEEE, 2012.
10. Q. Wang, Y. Guan, and J. Wang, "A chance-constrained two-stage stochastic program for unit commitment with uncertain wind power output," *IEEE Trans. Power Syst.*, vol. 27, no. 1, pp. 206–215, 2012.
11. A. Papavasiliou and S. Oren, "A comparative study of stochastic unit commitment and security-constrained unit commitment using high performance computing," in *Control Conference (ECC), 2013 European*, pp. 2507–2512, Jul. 2013.
12. D. Bertsimas, E. Litvinov, X. A. Sun, J. Zhao, and T. Zheng, "Adaptive robust optimization for the security constrained unit commitment problem," *IEEE Trans. Power Syst.*, vol. 28, no. 1, pp. 52–63, 2013.

13. C. Zhao and Y. Guan, "Unified stochastic and robust unit commitment," *IEEE Trans. Power Syst.*, vol. 28, pp. 3353–3361, Aug. 2013.
14. R. Jiang, J. Wang, M. Zhang, and Y. Guan, "Two-stage minimax regret robust unit commitment," *IEEE Trans. Power Syst.*, vol. 28, pp. 2271–2282, Aug. 2013.
15. R. Jiang, J. Wang, and Y. Guan, "Robust unit commitment with wind power and pumped storage hydro," *IEEE Trans. Power Syst.*, vol. 27, pp. 800–810, May 2012.
16. Q. Wang, J.-P. Watson, and Y. Guan, "Two-stage robust optimization for n-k contingency-constrained unit commitment," *IEEE Trans. Power Syst.*, vol. 28, pp. 2366–2375, Aug. 2013.
17. C. Zhao, J. Wang, J.-P. Watson, and Y. Guan, "Multi-stage robust unit commitment considering wind and demand response uncertainties," *IEEE Trans. Power Syst.*, vol. 28, pp. 2708–2717, Aug. 2013.
18. D. Mejia and J. McCalley, "Affine decision rules based power systems planning under multiple uncertainties," in *Presentation at the IEEE PES General Meeting*, Jul. 2011.